Advanced Materials for Sustainable Developments

Advanced Materials for Sustainable Developments

*A Collection of Papers Presented at the
34th International Conference on Advanced
Ceramics and Composites
January 24–29, 2010
Daytona Beach, Florida*

Edited by
Hua-Tay Lin
Andrew Gyekenyesi
Linan An

Volume Editors
Sanjay Mathur
Tatsuki Ohji

A John Wiley & Sons, Inc., Publication

Published by John Wiley & Sons, Inc., Hoboken, New Jersey.
Published simultaneously in Canada.

For general information on our other products and services or for technical support, please contact our
Customer Care Department within the United States at (800) 762-2974, outside the United States at
(317) 572-3993 or fax (317) 572-4002.

Wiley also publishes its books in a variety of electronic formats. Some content that appears in print may
not be available in electronic format. For information about Wiley products, visit our web site at
www.wiley.com.

Library of Congress Cataloging-in-Publication Data is available.

ISBN 978-0-470-59474-2

Printed in the United States of America.

10 9 8 7 6 5 4 3 2 1

Contents

Preface

Contributions from three symposia that were part of the 34th International Conference on Advanced Ceramics and Composites (ICACC), in Daytona Beach, FL, January 24-29, 2010 are presented in this volume. The broad range of topics is captured by the symposia titles, which are listed as follows: International Symposium on Ceramics for Electric Energy Generation, Storage, and Distribution (debuted in 2010); Thermal Management Materials and Technologies (debuted in 2010); and lastly, and Advanced Sensor Technology, Developments and Applications (debuted in 2010). These new symposia emerged during this ICACC meeting due to community growth and interest, and thus each of these subject areas were established as stand-alone symposia.

The current volume represents 15 contributions from the above listed symposia that embody the latest developments in engineering ceramics for energy technologies, thermal management utilizing either highly conductive or insulating materials, as well as advances regarding the utilization of ceramics for sensors.

The editors wish to thank the authors and presenters for their contributions, the symposium organizers for their time and labor, and all the reviewers for their valuable comments and suggestions. Acknowledgment is also due for the financial support from the Engineering Ceramic Division and The American Ceramic Society. Lastly, the editors would like to give thanks to the staff of the meetings and publication departments of The American Ceramic Society for their invaluable assistance.

HUA-TAY LIN, Oak Ridge National Laboratory
ANDREW GYEKENYESI, OAI, NASA Glenn Research Center
LINAN AN, University of Central Florida

Introduction

This CESP issue represents papers that were submitted and approved for the proceedings of the 34th International Conference on Advanced Ceramics and Composites (ICACC), held January 24–29, 2010 in Daytona Beach, Florida. ICACC is the most prominent international meeting in the area of advanced structural, functional, and nanoscopic ceramics, composites, and other emerging ceramic materials and technologies. This prestigious conference has been organized by The American Ceramic Society's (ACerS) Engineering Ceramics Division (ECD) since 1977.

The conference was organized into the following symposia and focused sessions:

Symposium 1	Mechanical Behavior and Performance of Ceramics and Composites
Symposium 2	Advanced Ceramic Coatings for Structural, Environmental, and Functional Applications
Symposium 3	7th International Symposium on Solid Oxide Fuel Cells (SOFC): Materials, Science, and Technology
Symposium 4	Armor Ceramics
Symposium 5	Next Generation Bioceramics
Symposium 6	International Symposium on Ceramics for Electric Energy Generation, Storage, and Distribution
Symposium 7	4th International Symposium on Nanostructured Materials and Nanocomposites: Development and Applications
Symposium 8	4th International Symposium on Advanced Processing and Manufacturing Technologies (APMT) for Structural and Multifunctional Materials and Systems
Symposium 9	Porous Ceramics: Novel Developments and Applications
Symposium 10	Thermal Management Materials and Technologies
Symposium 11	Advanced Sensor Technology, Developments and Applications

Focused Session 1 Geopolymers and other Inorganic Polymers
Focused Session 2 Global Mineral Resources for Strategic and Emerging
                   Technologies
Focused Session 3 Computational Design, Modeling, Simulation and
                   Characterization of Ceramics and Composites
Focused Session 4 Nanolaminated Ternary Carbides and Nitrides (MAX Phases)

The conference proceedings are published into 9 issues of the 2010 Ceramic Engineering and Science Proceedings (CESP); Volume 31, Issues 2–10, 2010 as outlined below:

- Mechanical Properties and Performance of Engineering Ceramics and Composites V, CESP Volume 31, Issue 2 (includes papers from Symposium 1)
- Advanced Ceramic Coatings and Interfaces V, Volume 31, Issue 3 (includes papers from Symposium 2)
- Advances in Solid Oxide Fuel Cells VI, CESP Volume 31, Issue 4 (includes papers from Symposium 3)
- Advances in Ceramic Armor VI, CESP Volume 31, Issue 5 (includes papers from Symposium 4)
- Advances in Bioceramics and Porous Ceramics III, CESP Volume 31, Issue 6 (includes papers from Symposia 5 and 9)
- Nanostructured Materials and Nanotechnology IV, CESP Volume 31, Issue 7 (includes papers from Symposium 7)
- Advanced Processing and Manufacturing Technologies for Structural and Multifunctional Materials IV, CESP Volume 31, Issue 8 (includes papers from Symposium 8)
- Advanced Materials for Sustainable Developments, CESP Volume 31, Issue 9 (includes papers from Symposia 6, 10, and 11)
- Strategic Materials and Computational Design, CESP Volume 31, Issue 10 (includes papers from Focused Sessions 1, 3 and 4)

The organization of the Daytona Beach meeting and the publication of these proceedings were possible thanks to the professional staff of ACerS and the tireless dedication of many ECD members. We would especially like to express our sincere thanks to the symposia organizers, session chairs, presenters and conference attendees, for their efforts and enthusiastic participation in the vibrant and cutting-edge conference.

ACerS and the ECD invite you to attend the 35th International Conference on Advanced Ceramics and Composites (http://www.ceramics.org/icacc-11) January 23–28, 2011 in Daytona Beach, Florida.

Sanjay Mathur and Tatsuki Ohji, Volume Editors
July 2010

DYE-SENSITIZED SOLAR CELL BASED ON ANODIC TiO₂ NANOTUBES PRODUCED FROM ANODIZATION IN FLUORIDE-FREE ELECTROLYTE

Narges F.Fahim and Tohru Sekino

Institute of Multidisciplinary Research for Advanced Materials, Tohoku University, Katahira 2-1-1, Aoba-ku, Sendai 980-8577, Miyagi Pref., Japan.

ABSTRACT

An increasing energy demand and environmental pollution create a pressing need for clean and sustainable energy solutions. TiO_2 semiconductor material is expected to play an important role in helping solve the energy crisis through effective utilization of solar energy based on photovoltaic devices. Dye-sensitized solar cells (DSSCs) are potentially lower cost alternative to inorganic silicon-based photovoltaic. In the present work, we report about the fabrication of dye-sensitized solar cell (DSSCs) from anodic TiO_2 nanotubes powder, produced by potentiostatic anodization of Ti foil in 0.1 M $HClO_4$ electrolyte, as photoanode. The counter electrode was made by electrodeposition of Pt from an aqueous solution of 5 mM H_2PtCl_6 onto fluorine-doped tin oxide glass substrate (FTO-glass). The above frontside illuminated DSSCs were compared with back-side-illuminated DSSCs fabricated from anodic TiO_2 NTs that were grown on the top of Ti foil as photoanode. The highest cell efficiency was 3.54 % under 100 mW/cm² light intensity (1 sun AM 1.5 G light, Jsc = 14.3 mA/cm², Voc = 0.544 V, fill factor = 0.455). To the best of our knowledge, this is the first report on the fabrication of dye-sensitized solar cell from anodic TiO_2 NTs powder. The TiO_2/FTO photoanodes were characterized by FE-SEM, XRD and Uv-visible spectroscopy. The catalytic properties of Pt/FTO counter electrodes have been examined by cyclic voltammetry.

1. INTRODUCTION

Dye-sensitized solar cells (DSSCs) are receiving an ever-increasing amount of attention as a promising photovoltaic technology, because of their potentially inexpensive manufacturing technology compared to silicon solar cells.[1] The front electrode of a DSSC is a transparent conductive oxide glass (TCO) coated with nanoporous TiO_2 covered with a monolayer of the Ruthenium complex based dye while the counter electrode is a TCO glass coated with platinum. The gap between the two electrodes is filled with an electrolyte containing an iodide / triiodide (I^-/I_3^-) redox couple. Upon illumination, the photoexcited dye injects an electron into the conduction band of the TiO_2. The electrons migrate through the TiO_2 network to be collected at the transparent counter electrode substrate. The light-sensitizing dye is subsequently regenerated electrochemically through an electrolyte. So far, the highest reported efficiency for a small area DSSC (< 0.2 cm²) is 11.3%, using acetonitrile electrolytes.[2]

TiO_2 nanotube arrays fabricated by anodization of titanium are now the focus of research in the fields of catalytic oxidation of organics[3], water photolysis[4], solar energy conversion[5], gas sensing[6], drug delivery[7], and biomedical implants[8,9] because of several attractive properties related to the unique material architecture of these nanotube (NT) arrays. For dye-sensitized solar cells, the properties of interest include the large internal surface area, enhanced optical absorption due to scattering effects[10], and lower recombination. However, most of the work has been focused on anodization of titanium foils to generate nanotube bundles on Ti surfaces. Application of such a material architecture in functional devices is limited due to the following: (a) the presence of an opaque Ti foil underneath a nanostructure in certain applications of optical-electrical devices (e.g., dye -sensitized solar cells) requires backside illumination that reduces the light-to-electrical energy conversion because of the platinized counter electrode partially reflects light and iodine in the electrolyte absorbs photons in the near UV region, and (b) the presence of a titanium metal underneath nanotubes may result in an electrical short-circuit when they are utilized in devices. Therefore, there is a pressing need to fabricate high aspect ratio TiO$_2$ nanotubes in a form that can be applied for a variety of functional substrates for various technological applications.

Our group was successfully prepared high aspect ratio TiO$_2$ nanotube powder by using rapid breakdown anodization method (RBA) [11]. It was reported that TiO$_2$ nanotubes produced by rapid breakdown anodization show a significantly higher photoresponse and conversion efficiencies than tubes that formed under self-ordering conditions[12]. Also, Chin-Jung et al.[13] studied the effect of anodic TiO$_2$ powder as additive in nanocrystalline TiO$_2$ film on electron transport properties and they observed an increase of more than 20 % in photocurrent density after addition of anodic TiO$_2$ powder. Moreover, Grimes and co-workers[14] anodized titanium thin films/ FTO glass and used them in a frontside dye-sensitized solar cell. They found conversion efficiencies of 2.9% from a 360-nm thick nanotube film. The method demonstrated by Grimes's group enables the front side illumination, but with a limited thickness (300-500 nm). By comparing our TiO$_2$ nanotube powders with freestanding TiO$_2$ nanotube membrane[15, 16] in terms of photovoltaics applications, the following can be withdrawn: first, the thickness of free-standing membrane of TiO$_2$ nanotube is usually in the range of 100-145 μm and such high thickness is not required for most applications, specially for photovoltaic solar cells[17] where the optimum thickness is in the range of 20-30 μm. Also, Jennings et al.[18] estimated the diffusion length of photoinjected electrons in TiO$_2$ nanotube cells and they found it to be on the order of 100 μm, which greatly influences the transport of electrons when TiO$_2$ NTs are used as photoactive anode in devices. By using high aspect ratio TiO$_2$ nanotube powders, which were prepared in previous study, the thickness can be controlled. Moreover, the production cost can be reduced by using TiO$_2$ NTs powders produced by RBA that formed in aqueous electrolyte in short time by a one-step method, in contrast to free-standing membrane of TiO$_2$ NTs, which requires long time (60 h) in nonaqueous electrolyte (ethylene glycol), and this issue is of significant importance for practical applications. Second, to use either TiO$_2$ NTs powders or free-standing TiO$_2$ NTs membrane in dye-sensitized solar cells, one should anchor the material to a supporting substrate such as FTO glass. Nanotube powders coated the conductive glass either by screen printing or by spin coating, while free-standing membrane can be anchored to conductive glass by using transparent conductive glue, which could increase the resistance of the cell and hence the photocurrent may be reduced. However, to the best of the authors' knowledge, up to now there have been no reports on the fabrication of photoanode from anodic TiO$_2$ nanotubes powder, produced by rapid breakdown anodization, as the active front-electrode in dye sensitized solar cell.

In the present work we describe a simple method for fabrication of dye-sensitized solar cells from anodic TiO$_2$ NTs powder as photoanode. The photovoltaic properties of these DSSCs were compared to the dye-sensitized solar cell fabricated by TiO$_2$ NTs that were grown on the top of Ti foil. It was found that solar cells fabricated using anodic TiO$_2$ nanotubes powder on FTO glass subjected to AM 1.5G (one sun) illumination in the frontside geometry exhibited an overall conversion efficiency of 3.54%. An efficiency of 0.88% was obtained in the back-side illumination geometry using anodic TiO$_2$ nanotubes on Ti foil. However, the photoanode fabricated from anodic TiO$_2$ nanotube powder was characterized by FE-SEM, XRD and Uv-visible spectroscopy. In addition, the catalytic properties of Pt counter electrodes were examined by cyclic voltammetry.

2. MATERIALS & METHODS

2.1. Materials

Ti foils (99.5% purity, 0.1-0.25 mm thickness) were purchased from Nilaco Corp. and, before anodization, cleaned by sonicating in acetone, 2-propanol, and methanol, followed by rinsing with deionized water and drying in a nitrogen stream. The anodization electrolyte was HClO$_4$ purchased from Sigma Aldrich, and all the solutions were prepared from reagent-grade chemicals and deionized water (Milli-Q). Lithium iodide (Wako), LiClO$_4$ (Aldrich), Iodine (Aldrich) were used as received. 4-*tert*-Butylpyridine (Wako), acetonitrile (Wako), and valeronitrile (Wako), Guanidinium thiocyanate (Aldrich) and H$_2$PtCl$_6$.6H$_2$O (Wako) were used without further purifications. TiCl$_4$ (Wako) was diluted with ethanol to 0.2 M at room temperature, which was kept in a refrigerator and freshly prepared for each TiCl$_4$ treatment of the TiO$_2$/FTO-coated glass photoelectrode. The dye N719 *cis*-di(thiocyanato)-*N, Ń* -bis(2,2′-bipyridyl-4-carboxylic acid-4′-tetrabutylammonium carboxylate) ruthenium (II) (N-719) was purchase from Dyesol Ltd., Australia. The dye solution was prepared by dissolving 0.1 g of dye (N-719) in 315 ml of dry ethanol then leave solution stirring overnight. 1-propyl-3-methylimidazolium iodide (PMII) (Merck). Anodic TiO$_2$ nanotubes powder and anodic TiO$_2$ nanotubes onto the surface of titanium foil were prepared as reported previously.[9, 19]

2.2. Preparation of photoanode

For the DSSC working electrodes, the FTO glass plates (1 mm thickness, 10 Ω/ square, Nippon Sheet Glass) were cleaned in an ethanol, acetone and iso-propanol using an ultrasonic bath for 15 min and then dried with N$_2$ gas stream. The FTO glass plates were soaked into aqueous solution of 0.2 M TiCl$_4$ for 16 h in airtight bottle, then washed with water and ethanol and finally dried and sintered at 450 °C for 30 min. Anodic TiO$_2$ nanotubes paste was prepared as follows: 6 g of anodic TiO$_2$ powder (annealed at 500 °C for 3 h in air) is ground in a porcelain mortar with a small amount of H$_2$O (2 ml) containing acetylacetone (0.2 ml) to prevent reaggregation of the particles. After the powder dispersed, it is diluted by slow addition of H$_2$O (8 ml) under continuous grinding. The paste of TiO$_2$ was homogenized for 30 min. Finally a detergent (0.1 ml Triton-100) is added to facilitate the spreading of the colloid on the substrate. The conducting FTO glass is covered on two parallel edges with adhesive tape to control the thickness of the TiO$_2$ layer. The anodic TiO$_2$ NTs paste is applied to one of the free edges of the substrate and distributed with a glass rod sliding over the tape-covered edges by doctor blade method. Keep the samples in a clean box for 3 min so that the paste could relax to reduce the surface irregularity and then dried for 3 min at 125 °C (at hot plate). This (Coating, storing and drying was repeated to change the thickness of the working electrode to 20 µm. finally, Sintering the photoanode (anodic TiO₂ paste/ FTO glass) at 450 °C for 1 h. The performance of the TiO₂ layer was improved by further deposition of TiO₂ from ethanolic solution of 0.2 M TiCl₄ by soaking the photoanode for 1 h at room temperature in airtight bottle. Then the photoanode was washed with dry ethanol. Immediately before being dipped into the dye solution, it was annealed again at 450 °C in air for 30 min then cooled to 80 °C. The resulting TiO₂ layer was sensitized by immersing it for 24 h in ~ 0.3 mM ethanolic solution of the N719 dye. After immersion, the photoanode was washed with anhydrous ethanol to remove the physisorbed dye. DSSC samples have been fabricated with active areas ranging in size from 0.4 to 0.16 cm².

For comparison, another kind of anodic TiO₂ NTs was prepared by anodization of titanium foil (0.25 mm thickness) in 0.1 M HClO₄ aqueous solution for about 10 sec. Before anodization, Ti foil was treated with HF/HNO₃ solution for 30 sec for uniform growth of TiO₂ NTs over all the surface of Ti foil.[19] The as-grown TiO₂ NTs/Ti foil was annealed at 500 °C for 3h in air to transfer the TiO₂ NTs layer from amorphous into crystalline structure to be photoactive. Then the photoanode was further treated with 0.2 M TiCl₄ ethanolic solution for 1h at room temperature. After treatment, it was washed with ethanol, then dried with N₂ gas and sintered at 450 °C for 30 min. After cooling to 80 °C, the TiO₂ electrode was immersed into the dye solution and kept at room temperature for 24 h to complete the sensitizer uptake. DSSCs fabricated from these photoanode was illuminated from back-side (pt-counter electrode).

2.3. Preparation of Pt-counter electrode

The electrodeposition experiments of Pt onto fluorine-doped tin oxide glass substrate (FTO-glass, 1.5 cm x 1.5 cm) were performed by using potentiostat/glavanostat (AMEL 7050, Italy). A two-electrodes cell was used with FTO-glass as cathode and a Pt foil as anode. The potentiostatic deposition of Pt film on the FTO substrate was carried out at a constant potential of 1.8 V in an electrolyte of 5 mM H₂PtCl₆.6H₂O aqueous solution for 1 min. The Pt-deposited FTO glass was dried at 100 °C for 10 min and annealed at 450 °C for 10 min. All the Pt counter electrode of the DSSCs used in this study had an area of 1 cm x 1 cm.

For thermal deposition, a drop of (5 mM H₂PtCl₆.6H₂O) in iso-propanol was placed on the FTO glass substrate, followed by annealing at 450 °C for 15 min. However, Pt counter electrode from (Dyesol Ltd., Australia) was used for comparison.

2.4. Fabrication of DSSCs devices

The dye-covered TiO₂ NTs electrode and Pt-counter electrode were assembled into a sandwich type cell spaced and sealed with a thermoplastic film of Surlyn of 30 µm thickness (Dyesol Ltd., Australia). A drop of the electrolyte, a solution of 0.60 M PMII, 0.03 M I₂, 0.10 M guanidinium thiocyanate and 0.50 M 4-tertbutylpyridine in a mixture of acetonitrile and valeronitrile (volume ratio: 85 : 15) was injected into the interelectrode space by the help of capillary action. Four kinds of DSSCs were prepared in this study as listed in Table I. Figure 1 shows a schematic diagram of frontside illuminated DSSC structure utilizing anodic TiO$_2$ nanotube powder paste and the digital photograph of the real cell.

2.5. Photocurrent-voltage measurements

The irradiation source for the photocurrent-voltage (I-V) measurement is a xenon light source (Model: YSS-50), which simulates the solar light. The incident light intensity was calibrated with a standard Si solar cell. Photovoltaic measurements employed an AM 1.5 solar simulator (100 mW/ cm^2) by applying an external bias to the cell and measuring the generated photocurrent with an Advantest R6245 (2) Channel Voltage/Current Source Monitor.

2.6. Characterization of DSSCs

TiO$_2$ NTs powder paste electrode and Pt counter electrode were initially characterized using a field emission SEM ((FESEM, Hitachi S-5000, Tokyo, Japan) at accelerating voltages of 20 kV. Crystallographic, the crystallinity and the purity information were recorded by X-ray diffraction using a powder X-ray diffractometer (Rigaku RINT 2500, Tokyo, Japan) with Cu Kα radiation (λ=1.54°A) at 40 kV and 50 mA with a scan rate of 0.02 °/s and a scan speed of 1°/min over a 2θ range from 10 to 80°. The resulting XRD spectra were compared with titania (anatase and rutile; JCPS No. 21-1272 and JCPS No. 21-1276, respectively). Diffuse-reflectance UV-visible spectra for evaluation of absorption properties of DSSC electrodes were recorded in the diffuse-reflectance mode (% R) and transformed to absorbance spectra using the Kubelka–Munk model.[20] The spectra were recorded on a Jasco-570 (Japan) spectrophotometer equipped with an integrating sphere; BaSO$_4$ was used as a standard.

The catalytic properties of Pt counter electrodes were determined by cyclic voltammetry. The measurements were performed in an electrolyte solution containing 1 mM I$_2$, 10 mM LiI and 0.1 M LiClO$_4$ in acetonitrile, at a scan rate of 100 mV/s, using potentiostat/glavanostat (AMEL 7050, Italy). The electrochemical cell consisted of Pt working electrode (1cm x 1cm), prepared by electrochemical deposition, thermal and Pt counter electrode from Dyesol, a pt foil auxiliary electrode and an Ag/AgCl reference electrode. The presence of LiClO$_4$ suppress the electrostatic migration of electroactive species in the electrochemical cell, which is absent in a DSSC.

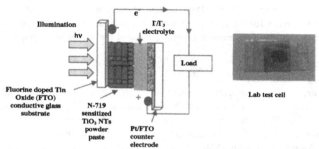

Figure 1. Schematic representation of the architecture used to Fabricate DSSC device utilizing the anodic TiO$_2$ nanotube powder paste (frontside illuminated).

3. RESULTS & DISCUSSION

3.1. Morphological characterization of DSSC photoanode

Figure 2 illustrates the field emission scanning electron microscope images, which show the morphology of the photoanode made from anodic titania NTs powder by forming a paste and applied on FTO glass by doctor blade method. As can be seen, the anodic TiO$_2$ NTs is good adhering to each other and there are no agglomerates. The thickness of the TiO$_2$ layer is about 20 μm, as measured by SEM. From our previous study, [11] the wall thickness and the outer diameter of TiO$_2$ NTs power were 8-15 nm and 40 nm, respectively. Moreover, figure 3c shows the high resolution transmission electron microscope of the anodic titania nanotubes powder. Obviously, the nanotubular structure can be seen.

3.2. Crystal structure of anodic TiO$_2$ photoanode

Because the anodic TiO$_2$ nanotube powder has been annealed at 500 °C for 3 h, the nanotubes have been converted to crystalline titania in the anatase phase (major) and rutile (minor). This is demonstrated in Figure 3, which shows the X-ray diffraction pattern of the anodic TiO$_2$ nanotube powder paste on FTO glass substrate. The spectrum consists of a variety of peaks that are completely attributable to anatase TiO$_2$ and rutile, as indicated. The presence of the peak at 25.45° (anatase) and at 27.45° (rutile) indicates that the nanotubes consist entirely of the anatase phase and rutile. Also, the XRD pattern of FTO glass substrate without anodic TiO$_2$ powder paste was included for comparison.

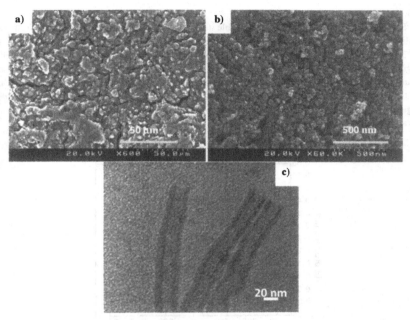

Figure 2. FE-SEM micrographs of photo-anode composed of anodic TiO₂ NTs powder paste/FTO glass (a,b) and HRTEM image of TiO₂ NTs powder (c).

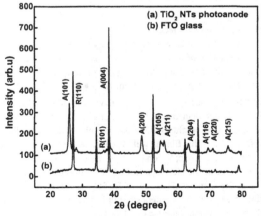

Figure 3. XRD patterns of anodic TiO₂ NTs/FTO electrode and FTO glass substrate.

3.3. Optical absorption properties

Figure 4 compares the ultraviolet-visible (UV-vis) absorption spectra obtained from diffuse-reflactance measurements of 20 μm anodic TiO$_2$ nanotube paste/FTO electrodes, pristine and coated with N-719 dye as well as Pt counter electrode and FTO glass substrate. It is clearly evident from Figure 4 that the anodic TiO$_2$ NT electrode sensitized with N-719 dye has higher optical absorption than the pristine anodic TiO$_2$ NTs electrode (without dye) across the entire solar spectrum. It is clear that coating the anodic TiO$_2$ NTs with dye greatly enhances the optical absorption at the solar spectrum region. However, the pristine anodic TiO$_2$ electrode has maximum absorption in the UV region and has no absorption in the wavelength region from 400 to 700 nm.

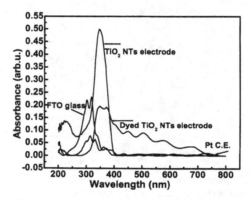

Figure 4. Diffuse-Reflectance Uv-vis spectra of anodic TiO$_2$ NTs electrode with and without dye sensitization as well as Pt counter electrode and FTO glass substrate.

3.4. Catalytic properties of Pt counter electrodes

The active surface areas of the Pt films were confirmed by cyclic voltammetry; from the rates of the oxidation of iodide and the reduction of triiodide. Figure 5a shows the cyclic voltammograms of the Pt films. The anodic peak are denoted by I and the cathodic peak by I'. In an anodic sweep, iodide is oxidized to triiodide (peak I) according to this reaction: $3I^- \rightarrow I_3^- + 2e^-$. When the potential scan is reversed, triiodide is reduced to iodide (peak I'). Obviously, the anodic and cathodic peaks associated with electrodeposited Pt are larger than that for the thermal deposited Pt and Pt from Dyesol Ltd., Australia. The increased in the peak heights and areas of the electrodeposited Pt film can be explained in terms of increased surface area of this film as well as enhanced reduction of I$_3$ and oxidation of I$^-$ at the electrodeposited pt working electrode, [21] compared to the pt films obtained by the other methods. The increased active surface area of the electrodeposited Pt films is desirable for DSSC because of the fact that reduction of I$_3$ at the counter electrode of DSSC requires larger active surface area. The enhancement of the surface area of the electrodeposited Pt is favorable for the reduction of the I$_3$ at the counter electrode, which in turn enhances the Jsc of a DSSC.

Figure 5b shows the FE-SEM micrograph of the electrodeposited Pt films/FTO glass as the DSSC counter electrode, which represents the morphological properties of the pt films obtained by electrochemical deposition.

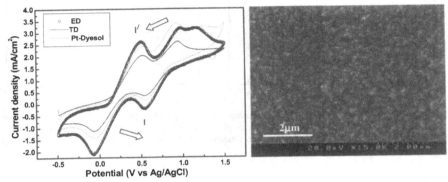

Figure 5. Cyclic voltammograms of the electrodeposited Pt films compared with thermal deposited Pt films and Pt counter electrode from Dyesol. (measurement conditions was performed in an electrolyte containing 1mM I_2, 10 mM LiI, 0.1M $LiClO_4$ in acetonitril at scan rate of 100 mV/s, the active surface area is 1 cm^2), (b) FE-SEM micrograph of electrodeposited Pt/FTO DSSC counter electrode.

Table I. list of DSSCs used in this study.

DSSCs	Photoanode	Pt-counter electrode	Treatment with TiCl₄ solution
DSSC1	TiO₂ NTs powder paste	Electrodeposited Pt	Treated
DSSC2	TiO₂ NTs powder paste	Electrodeposited Pt	None
DSSC3	TiO₂ NTs powder paste	Dyesol Pt counter electrode	Treated
DSSC4	TiO₂ NTs/Ti-foil annealed at 500 °C for 3 h in air.	Electrodeposited Pt	Treated

Table II. Photovoltaic characteristics of DSSCs based on anodic TiO₂ NTs produced from rapid breakdown anodization.

DSSC	Jsc (mA/cm²)	Voc (V)	FF	η (%)
DSSC1	14.3	0.544	0.455	**3.54**
DSSC2	4.95	0.576	0.538	1.53
DSSC3	7.9	0.613	0.492	3.41

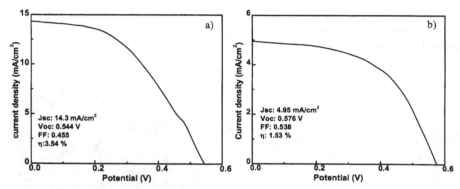

Figure 6. I-V characteristics of DSSCs using TiO₂ NTs powder, fabricated by rapid breakdown anodization in 0.1 M HClO₄ electrolyte, as photoanode: (a) treated with TiCl₄ solution and (b) untreated with TiCl₄ solution. (Frontside illumination cell geometry under standard AM 1.5 G illumination (100 mW/cm²).

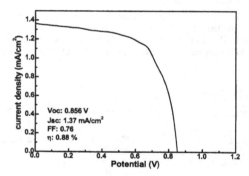

Figure 7. Photourrent density-voltge (I-V) characteristics of DSSC using TiO₂ NTs grew on titanium foil by rapid breakdown anodization in 0.1 M HClO₄ electrolyte and annealed at 500 °C. Backside illumination cell geometry under AM 1.5G illumination (100 mW/cm²).

3.5. Photovoltaic performance of DSSCs

Figure 6a shows the I-V characteristics of solar cells fabricated using 20 μm long anodic TiO₂ nanotube powder layers that are treated with TiCl₄ in comparison to nontreated layers. A cell fabricated from treated nanotubes showed a short-circuit photocurrent density (Jsc) of 14.3 mA/cm², an open circuit potential (Voc) of 0.544 V, and a fill factor of 0.455 resulting in an overall power conversion efficiency of 3.54 %. Untreated TiO₂ nanotubes were used to fabricate solar cells, whose photocurrent density-voltage (I-V) curve are shown in figure 6b, which exhibits a Jsc of 4.95 mA/cm^2, a Voc value of 0.576 V and a fill factor (FF) of 0.538 to produce an overall power conversion efficiency of 1.53%. An increase (> two times) can be seen in solar cell performance, which originates from the increase in current density. We believe that the TiCl$_4$ treatment improves performance by increasing the effective surface area available for dye adsorption by increasing the roughness of the tube walls. Further, the TiCl$_4$ treatment seems to improve bonding between the TiO$_2$ and the dye molecules, resulting in better charge injection.[22] It has been reported that titania nanotubes can efficiently adsorb large quantities of dye in a similar manner as the standard mesoporous nanostructured TiO$_2$ films deposited on conductive glass or foil.[23] On such porous, rough and fractal nanotubes the real surface is sufficiently large to allow the chemisorbed dye molecules to achieve relatively high concentration and therefore efficiently absorb and convert the sunlight into electricity.[24]

The above efficiency (3.54 %) was slightly decreased with the use of pt counter electrode from (Dyesol Ltd., Australia), reaching 3.41%, with Jsc=7.9 mA/cm^2, Voc=0.613 V and FF = 0.492 (the curve not shown). This decrease in the device performance can be understood from the cyclic voltammogram curve, where in case of electrodeposited pt, the peak of oxidation and reduction is larger than that of pt films obtained either by thermal deposition or from Dyesol. The parameters derived from the I-V characteristic of all DSSCs used in this study are listed in Table II. It is interesting to note that the efficiency of DSSC fabricated from TiO$_2$ nanotube arrays, obtained by alumina templating method [25], is 3.5 %, which is very similar to our DSSC efficiency obtained in this study.

Representative current-voltage (I-V) measurements are shown in Figure 7 for a DSSC fabricated from anodic TiO$_2$ nanotube arrays grown onto Ti foil. The active area of the DSSC was 0.196 cm^2. Upon backside illumination under AM1.5G conditions, the device exhibited an open-circuit voltage (Voc) of 0.856 V, a short-circuit current density (Jsc) of 1.37 mA/cm^2, and a fill factor (FF) of 0.76, giving an overall conversion efficiency (η) of 0.88 %. The strong reduction in the efficiency of this cell can be explained as follows: when using the backside configuration there are strong absorption from both the counter electrode and the electrolyte. On taking into account the presence of high quantity of iodine (M =0.03 M) absorbs a significant portion of the incoming light. One should note that Grätzel's pioneer DSSCs lose about 40 % of their efficiency when illuminated from the backside illumination (Pt counter electrode). [26] The above efficiency is as high as that reported by Han et al.[12] for DSSC using RBA-based tubes (0.63 %). It is noteworthy that the corresponding DSSCs fabricated using nanoparticulate opaque TiO$_2$ thin films (about 10–15 µm in thickness) electrodes led to relatively lower efficiencies (Jsc = 1.43 mA/cm^2, Voc = 0.644 V, FF = 0.59, η = 0.54%). This direct comparison is very encouraging towards application of TiO$_2$ nanotubes in solar cell configurations.

Clearly, the fabrication of DSSC from anodic TiO$_2$ nanotube in the powder form promotes the frontside illumination configuration. As a consequence these materials enhance the overall conversion efficiency by a factor of four than the DSSC fabricated using nanotube/Ti foil.

Nevertheless, the high significance of the results obtained using the anodic TiO$_2$ nanotubes in the powder form has been established, since we achieved an increase (by four times) of the efficiency reported for randomly ordered RBA tubes grew onto the surface of Ti foil.

4. CONCLUSIONS

High-efficiency dye-sensitized solar cells were fabricated from anodic TiO_2 nanotube arrays, produced by rapid breakdown anodization for photoanode. Two kinds of anodic TiO_2 nanotubes have been used one is anodic TiO_2 nanotube powder and the other is anodic TiO_2/Ti foil. Pt- counter electrodes were prepared by different methods onto FTO glass. These devices were characterized by FE-SEM, Uv-visible spectroscopy, cyclic voltammetry and photocurrent-voltage measurements. Dye-sensitized solar cells fabricated using $TiCl_4$-treated anodic TiO_2 nanotubes powder paste layers as working electrode yielded power conversion efficiency as high as 3.54 % under 100 mW/cm² (AM 1.5 G) frontside illumination. These results suggest a promising approach to achieve high efficiency devices. The RBA titania nanotubes grown in a very short time are a viable candidate for the fabrication of DSSCs, in spite of their less ordered structure. Our results are of importance to realize the application of anodic TiO_2 in fabrication of DSSC and to provide useful clues for further improvement in the device efficiency.

REFERENCES
[1] B. O' Regan and M. Grätzel, A Low-Cost, High-Efficiency Solar Cell Based on Dye-Sensitized Colloidal TiO₂ Films, Nature **353** 737-40 (1991).

[2] M. Grätzel, The Advent of Mesoscopic Injection Solar Cells, Prog. Photovolt. **14**, 429-42 (2006).

[3] Z. Liu, X. Zhang, S. Nishimoto, M. Jin, D. A.Tryk, T. Murakami and A. Fujishima, Highly Ordered TiO₂ Nanotube Arrays with Controllable Length for Photoelectrocatalytic Degradation of Phenol, J. Phys. Chem. C **112**, 253-59(2008).

[4] G. K.Mor, H. E. Prakasam, O. K. Varghese, K. Shankar and C. A. Grimes, Vertically Oriented Ti–Fe–O Nanotube Array Films: Toward a Useful Material Architecture for Solar Spectrum Water Photoelectrolysis, Nano Lett. **7**, 2356-364 (2007).

[5] K.Shankar, G. K. Mor, H. E. Prakasam, O. K. Varghese, C. A. Grimes, Self-Assembled Hybrid Polymer–TiO₂ Nanotube Array Heterojunction Solar Cells, *Langmuir*, **23**, 12445- 2449(2007).

[6] O. K. Varghese, X. Yang, J. Kendig, M. Paulose, K. Zeng, C. Palmer, K. G. Ong and C. A. Grimes, A Transcutaneous Hydrogen Sensor: From Design to Application , *Sensor Letters*, **4**, 120- 128 (2006).

[7] K. C. Popat, M. Eltgroth, T. J. La Tempa, C. A. Grimes, T. A. Desai, Titania Nanotubes: A Novel Platform for Drug-Eluting Coatings for Medical Implants, *Small*, **3**, 1878- 881 (2007).

[8] K. C. Popat, L. Leoni, C. A. Grimes, T. A. Desai, Influence of engineered titania nanotubular surfaces on bone cells. Biomaterials, **28**, 3188-97 (2007).

[9] N.F.fahim, M.F. Morks and T. Sekino, Electrochemical Synthesis of Silica-Doped High Aspect-Ratio Titania Nanotubes as Nanobioceramics for Implant Applications, Electrochim. Acta **54**, 3255–269 (2009).

[10] K. G. Ong, O. K. Varghese, G. K. Mor, K. Shankar, C. A. Grimes, Application of Finite-Difference Time Domain to Dye-Sensitized Solar Cells: The Effect of Nanotube-Array Negative Electrode Dimensions on Light Absorption, Solar Energy Mater & Solar Cells, **91**, 250-57 (2007).

[11] N. F. Fahim and T. Sekino, A Novel Method for Synthesis of Titania Nanotube Powders using Rapid Breakdown Anodization, Chem. Mater., **21**, 1967–79 (2009).

[12] R. Hahn, T. Stergiopoulus, J. M. Macak, D. Tsoukleris, A. G. Kontos, S. P. Albu, D. Kim, A. Ghicov, J. Kunze, P. Falaras, P. Schmuki , Efficient solar energy conversion using TiO$_2$ nanotubes produced by rapid breakdown anodization - a comparison, phys. stat. sol. (RRL), **1**, 135-137 (2007).

[13] C-J Lin, W-Y Yu, and S.H. Chien, Effect of Anodic TiO$_2$ Powder as Additive on Electron Transport Properties in Nanocrystalline TiO$_2$ Dye-Sensitized Solar Cells, Appl. Phys. Lett., **91**,233120-23 (2007).

[14] G. K. Mor, K. Shankar, M. Paulose, O. K. Varghese and C. A. Grimes, Use of Highly-Ordered TiO$_2$ Nanotube Arrays in Dye-Sensitized Solar Cells, Nano Lett., **6**, 215-18 (2006).

[15] J. Wang and Z. Lin, Freestanding TiO$_2$ Nanotube Arrays with Ultrahigh Aspect Ratio via Electrochemical Anodization, Chem. Mater., **20**, 1257-61 (2008).

[16] S. Albu, A.Ghicov, J. Macak, R. Hahn, P. Schmuki, Self-Organized, Free-Standing TiO$_2$ Nanotube Membrane for Flow-through Photocatalytic Applications, Nano Lett., **7**, 1286-89 (2007).

[17] J.M. Macak, H. Tsuchiya, A. Ghicov, K. Yasuda, R. Hahn, S. Bauer, P. Schmuki, TiO$_2$ Nanotubes: Self-Organized Electrochemical Formation, Properties and Applications, Curr. Opin. Solid State Mater. Sci., **11**, 3-18 (2007).

[18] J. R. Jennings, A. Ghicov, L. M. Peter, P. Schmuki, A. B. *J.* Walker, Dye-Sensitized Solar Cells Based on Oriented TiO$_2$ Nanotube Arrays: Transport, Trapping, and Transfer of Electrons, Am. Chem. Soc., **130**, 13364-72 (2008).

[19] N. F. Fahim, T. Sekino, M. F. Morks and Takafumi Kusunose, Electrochemical Growth of Vertically-Oriented High Aspect Ratio Titania Nanotubes by Rapid Anodization in Fluoride-Free Media, NanoScience & NanoTechnology, **9,** 1803-18 (2009).

[20] Z.C. Orel, M.K. Gunde, B. Orel, Application of the Kubelka-Munk Theory for the Determination of the Optical Properties of Solar Absorbing Paints, Progress in Organic Coatings, **30**, 59- 66(1997).

[21] K.S. Choi, E.W. Mc Farland,G. D. stucky, Electrocatalytic Properties of Thin Mesoporous Platinum Films Synthesized Utilizing Potential-Controlled Surfactant Assembly, Adv. Mater. **15**, 2018-21 (2003).

[22] P. M. Sommeling, B. C. O'Regan, R. R. Haswell, H. J. P. Smit, N. J. Bakker, J. J. T. Smits, J. M. Kroon, and J. A. M. van Roosmalen, Influence of a TiCl$_4$ Post-Treatment on Nanocrystalline TiO$_2$ Films in Dye-Sensitized Solar Cells, J. Phys. Chem. B **110**, 19191-97(2006).

[23] K. Zhu, N. R. Neale, A. Miedaner and A. J. Frank, Enhanced Charge-Collection Efficiencies and Light Scattering in Dye-Sensitized Solar Cells Using Oriented TiO$_2$ Nanotubes Arrays ,Nano Lett. **7**, 69- 74 (2007).

[24] P. Falaras, Synergetic Effect of Carboxylic Acid Functional Groups and Fractal Surface Characteristics for Efficient Dye Sensitization of Titanium Oxide, Sol. Energy Mater. Sol. Cells **53**, 163-75 (1998).

[25] T-S Kang, A. P. Smith, B. E. Taylor and M. F. Durstock, Fabrication of Highly-Ordered TiO$_2$ Nanotube Arrays and Their Use in Dye-Sensitized Solar Cells, Nano Lett., **9**, 601–606 (2009).

[26] M. Paulose, K. Shankar, O. K. Varghese, G. K. Mor, B. Hardin and C. A. Grimes, Backside Illuminated Dye-Sensitized Solar Cells based on Titania Nanotube Array Electrodes, Nanotechn. **17**, 1446- 48 (2006).

SELF-PROPAGATING HIGH-TEMPERATURE SYNTHESIS OF CALCIUM COBALTATE THERMOELECTRIC POWDERS

Sidney Lin and Jiri Selig
Dan F. Smith Department of Chemical Engineering
Lamar University
Beaumont, TX 77710
U.S.A.

Hua-Tay Lin and Hsin Wang
Materials Science and Technology Division
Oak Ridge National Laboratory
Oak Ridge, TN 37831
U.S.A.

ABSTRACT

Thermoelectric $Ca_{1.24}Co_{1.62}O_{3.86}$ powders are synthesized by Self-propagating High-temperature Synthesis (SHS), a simple and economical process of synthesis oxides. Prepared powders were analyzed by XRD for phase purity, TG/DSC for thermal stability and SEM for particle size. Post treatment is also applied to improve the phase purity of powders produced by SHS. A short post treatment of 15 minutes in air can convert the SHS product into phase pure $Ca_{1.24}Co_{1.62}O_{3.86}$. To study the reaction mechanism, TG-DSC analysis was performed on the reaction mixture to establish the reaction mechanism. Prepared powders have high Seebeck coefficient and figure of merit close to those prepared by spark plasma sintering. A finite element mathematical model is used to simulate the temperature profile inside the reactant pellet during the SHS.

INTRODUCTION

In an automobile, only 27% of the fuel energy is converted into kinetic energy, 33% used to cool the engine, 4% is lost as friction, and 36% of the fuel energy is lost as exhaust. The energy lost from the exhaust pipe is so large that it is estimated recycling 1% of the exhaust waste heat can power all accessories except the headlights in an automobile[1]. The US industry sector and the transportation sector consumed 32% and 29% of the national energy in 2007[2]. Fifty percent of this 32.4 Quad Btu was primary energy consumption of oil and gas to produce heat, steam, and cogeneration for the industry sector. At an estimation of a 60-80% process efficiency and a 2% thermoelectric system efficiency, 0.75-1.5 Megawatt of energy can be saved annually in the US by thermoelectric power generation.

Thermoelectric materials are compounds that can utilize temperature difference to generate electrical current or inversely convert electrical current into temperature difference. The Seebeck effect describes the generation of an electrical current in a closed loop when there is a temperature difference between two leads[3]. It results from the electrons in the high temperature zone vibrate and migrate faster than those in a cooler zone and generates net electrical current and potential. The opposite effect known as the Peltier effect describes the temperature difference presence when a current is passed through a closed loop. In recent years, the Seebeck effect is used to generate electrical powder from automobile exhaust heat and the Peltier effect is used at the same time to cool down the temperature of a car seat in the summer.

Thermoelectric materials are characterized by their thermoelectric figures of merit, *ZT*, defined by equation (1):

$$ZT = \frac{\alpha^2 \sigma}{\kappa} T \qquad (1)$$

where α is Seebeck coefficient, σ is electrical conductivity, T is temperature, and κ is thermal conductivity. From equation (1) it can be seen that a good thermoelectric material should have (i) a high Seebeck coefficient to produce a substantial voltage, (ii) a high electrical conductivity to minimize the Joule heating, and (iii) a low thermal conductivity to keep a high temperature difference within the material. However, these properties are closely related. Optimizing one property could degrade another one and yield no or negative net effect.

Equation (1) also implies that a viable thermoelectric compound has to be both electron conductor and heat insulator. Metal alloys, such as PbTe, SiGe, and Bi_2Te_3 were first studied for thermoelectrics because of their high electrical conductivities. However, these alloys have a tendency to oxidize at high temperatures and are not stable in an oxidizing atmosphere, such as inside the vehicle exhaust lines where temperature can reach 800 °C.

Oxides were thought to have poor thermoelectric properties because of their low electrical conductivities at room temperature. However, with the discovery of the thermoelectric properties of $NaCo_2O_4$ in 1997[4], oxide thermoelectric materials have been studied extensively. Complex calcium cobalt oxides have been reported with good thermoelectric properties, especially at high temperatures. They are good thermoelectric materials due to their layered structure, combining high mobility layers and phonon-glass layers[5]. All cobaltates share the similar structure; alternating ordered metallic CoO_2 blocks and insulating disordered blocks[6]. In the case of $Ca_{1.24}Co_{1.62}O_{3.86}$, $[CoO_2]$ blocks are alternating with rocksalt-type $[Ca_2CoO_3]_{0.62}$ blocks[7]. This mismatched structure leads to a reduced thermal conductivity of layered cobaltates due to high phonon scattering at the disordered boundaries of the different building blocks[8].

Traditionally complex calcium cobalt oxides are produced either by sintering metal oxides or carbonates at high temperatures for a long period of time[9] or by wet-chemical methods followed by sintering at high temperatures[10, 11]. These processes are energy intensive and not economical. A possible solution to make the production of thermoelectric oxides more economical is the use of Self-propagating High-temperature Synthesis.

Self-propagating High-temperature Synthesis (SHS) was developed in Russia in the late 1960's and has been used to synthesize many ceramic materials including oxides, nitrides, carbides, and metal hydrides. This process is highly exothermic. Only a small amount of ignition energy is needed to initiate the reaction. During the synthesis the heat generated from the reaction is sufficient to sustain the reaction progress. SHS process is usually conducted at room temperature, which makes the synthesis not energy intensive. The fast combustion front movement (1-100 mm/sec) enables a large-scale production in a short period of time. In addition, its fast cooling after the reaction allows the formation of ceramic powders with very fine grains and metastable compositions. The motivation of this work is to use SHS to produce nano structured calcium cobalt thermoelectric oxides economically.

EXPERIMENTAL

Stoichiometric amount of reactants shown in reaction (2) were first mixed in a jar mill for four hours and then pressed into pellets 7/8 inch in diameter and approximately 3 cm long. The prepared pellets were placed into a reactor with flowing oxygen and ignited by a graphite igniter which is controlled by a variable transformer. Temperature history of SHS reaction was measured by inserting two K-type thermocouples at a certain distance apart.

$$1.24\ CaO_2 + (1.62 - 3x)\ Co + x\ Co_3O_4 + O_2 \rightarrow Ca_{1.24}Co_{1.62}O_{3.86} \qquad (2)$$

Some of the calcium-cobalt oxide prepared by SHS was post treated in a tube furnace at 850 °C to study the effects of post treatment on product purity.

Reactants and SHS product were analyzed by TG/DSC (Netzsch STA 449C and DMA 242C) to determine the reaction scheme and thermal stability. Thermal analysis was performed in oxygen, nitrogen, and air with heating rates of 5 °C/min or 50 °C/min. X-ray diffraction (Bruker D8 Discover GADDS) was used to determine the phase purity of prepared powders. SEM (Hitachi S3400N with EDAX) was used to determine the particle size and shape. Thermoelectric properties were measured at Oak Ridge National Laboratory during a feasibility test. Samples were hot pressed at 850 °C under a load of 20 MPa. The Seebeck coefficient and resistivity were measured using ULVAC ZEM-3 system.

MATHEMATICAL MODEL

The mathematical model was developed based on our experimental set up. It consisted of a 3 cm long pellet with a 7/8" in diameter and a 10 cm long quartz sample holder underneath the sample pellet. Flowing oxygen surrounded the pellet and holder. Since oxygen flew into the reactor at a very low velocity (0.034 mm/s) and the reactor exit is open to the atmosphere ($p_{out} = 1$ atm) a laminar flow pattern was assumed. A weakly compressible Navier-Stokes equation was used due to the sharp temperature gradients during SHS reaction, which affects properties of oxygen and oxygen flow pattern.

SHS reaction was modeled in one step since the conversion from reactants to product occurs in few seconds. Reaction rate and heat generation were expressed by the following equations[12]:

$$\frac{d\eta}{dt} = k \cdot \exp\left(\frac{-E}{RT}\right)(1-\eta)^n \qquad (3)$$

$$q = \Delta H_r \frac{d\eta}{dt} \qquad (4)$$

where η is conversion, t is time, R is gas constant, T is temperature, and ΔH_r is enthalpy of reaction. Activation energy (E), was determined from experimental data, reaction order (n) was equal to 1 and pre-exponent (k) was equal to 1,000. Above equations were combined with overall energy balance. Inlet boundary condition was set to constant temperature (25 °C) and outlet was set to convective flux (common boundary for a system with unknown outlet temperature). Interior boundary conditions took into account included the convective and radiation heat losses.

RESLUTS AND DISCUSSIONS

SHS and Post Treatment

The pellet was ignited by a short heating without any difficulty. During the SHS reaction, the color of pressed pellet changed from light gray to dark gray. Water vapor evolution was observed, which was caused by calcium hydroxide impurity present in the reactant mixture. Figure 1 shows the temperature history at the center of the reactant pellet. In this measurement, two K-type thermocouples 1.7 cm apart from each other were placed at the centerline of the pellet. Form this figure, the speed of reaction front movement can be calculated to be 0.5 mm/s, which is relatively slow, compared to most SHS systems. The temperature at the centerline of the pellet increases from room temperature to a maximum temperature of 1,200 °C in less than 10 second and decreases below 850 °C in about 100 second. The reaction time (the time for the pellet temperature being above 850 °C) is around 150 seconds.

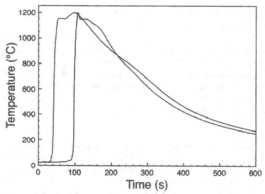

Figure 1. Temperature history at the center of the pellet during SHS of $Ca_{1.24}Co_{1.62}O_{3.86}$.

X-ray diffraction of SHS'ed sample (Figure 2) shows the formation of an unknown intermediate compound (not the desired thermoelectric phase), and no starting material was detected in the SHS product. The unknown peaks are believed to be either a calcium-cobalt oxide of unknown stoichometry or $Ca_{1.24}Co_{1.62}O_{3.86}$ of different lattice constants. We believe due to the fast SHS reaction and short reaction time (about 100 seconds for temperature being above 850 °C), reactants did not have enough time to form desired product (thermodynamic preferred one), but instead a metastable compound (kinetically preferred one) was formed.

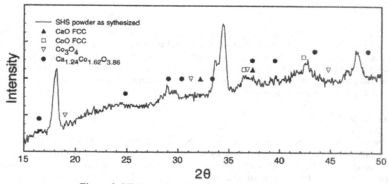

Figure 2. XRD pattern of powder prepared by SHS.

Figure 3 shows the SEM image of hand ground SHS powders. The average particle size is about 1 μm. The rough appearing of the particles in SEM image suggests a high surface area.

Figure 3. SEM of powder prepared by SHS.

To convert the intermediate compounds produced by SHS reaction to a thermodynamically preferred compound, a post heat treatment of the SHS'ed sample was studied by placing the SHS'ed powders inside a tube furnace at 850 °C. The results are shown in Figure 4. It can be seen that the desired thermoelectric compound ($Ca_{1.24}Co_{1.62}O_{3.86}$) peak was observed after a 15-minute post treatment at 850 °C in air. Increasing post treatment time did not have noticeable effect on product purity. From Figure 4, we can conclude that a 15 minute post treatment following the SHS reaction can produce phase pure $Ca_{1.24}Co_{1.62}O_{3.86}$. This short production time enables a large production of thermoelectric $Ca_{1.24}Co_{1.62}O_{3.86}$ in a shout period time.

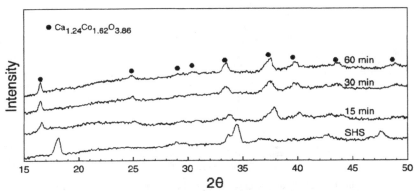

Figure 4. XRD patterns of powder prepared by SHS followed by post treatment in air at 850 °C for different times.

Figure 5 show the SEM image of a sample post treated for 30 minutes in air at 850 °C and subsequently hand-ground. Particle size varies from several microns to sub-micron.

Figure 5. SEM of powder prepared by SHS followed by post treatment in air at 850 °C for 15 minutes.

Thermal Analysis

A mixture of CaO_2 and Co (Ca:Co = 1.24:1.62) heated at a rate of 50 °C/min to simulate the fast temperature increase during the SHS. Three main reaction steps are observed from the thermal analysis (Fig. 6). The first one is the decomposition of CaO_2 to form CaO and O_2 between 400 and 450 °C. The second step is Co oxidation to form cobalt oxides and reaction between CaO and cobalt oxides to form $Ca_{1.24}Co_{1.62}O_{3.86}$, which start at 450 °C and continues to 1,000 °C. Third step is the decomposition of $Ca_{1.24}Co_{1.62}O_{3.86}/Co_3O_4$ mixture at 1,000 °C to form CaO, CoO and O_2.

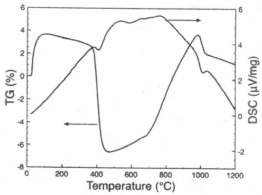

Figure 6. TG/DSC of mixture of CaO_2 and Co in O_2 heated at 50 °C/min

If all reactants were converted to $Ca_{1.24}Co_{1.62}O_{3.86}$, a 10% increase in mass should be observed. However, measured mass gain was only 4% which suggests an incomplete oxidation and product formation due to the fast heating rate. The product at the highest point of the TG curve is most likely a mixture of $Ca_{1.24}Co_{1.62}O_{3.86}$, CaO, and Co_3O_4. TG results of another sample heated at a slower heating rate of 5 °C/min (Fig 7.) showed a much larger mass increase. It suggests that in order to achieve a complete conversion a slow heating rate or a longer reaction/post treatment time is necessary.

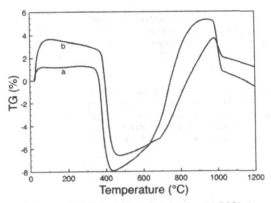

Figure 7. TG curve of mixture of CaO_2 and Co in O_2 heated at (a) 5 °C/min and (b) 50 °C/min.

Thermal analysis was also conducted to study the thermal stability of pure $Ca_{1.24}Co_{1.62}O_{3.86}$ prepared by SHS followed by post treatment (Fig. 8) in different atmospheres (N_2, O_2, and air). Blank test of alumina, which was placed at the bottom of sample holder to prevent possible reaction of sample with crucible is also shown in this figure. At sufficiently high temperatures $Ca_{1.24}Co_{1.62}O_{3.86}$ decomposes to CoO, CaO, and oxygen gas. The evolution of oxygen results a 7.7 % mass loss. For sample analyzed in N_2 the mass drop (weight changes from 79 % to about 72%) caused by decomposition occurs in two steps. The first step starts at 850 °C and the second step begins at 920 °C and completes at 1,000 °C. For sample in air mass drops (weight changes from 71.5 % to about 64 %) from decomposition occurs also in two very distinct steps. Decomposition starts at 950 °C and is complete at 1,100 °C. Sample analyzed in O_2 has a mass loss from 62.5 % to 53 %. Decomposition starts at 1,020 °C and is complete at 1,200 °C. Decomposition for of $Ca_{1.24}Co_{1.62}O_{3.86}$ in all three gases resulted in 7 to 9 % confirming complete decomposition.

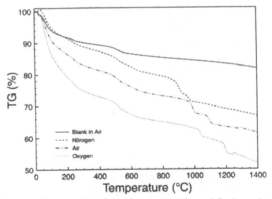

Figure 8. TG curves for pure $Ca_{1.24}Co_{1.62}O_{3.86}$ in N_2, Air, and O_2, heated at 5 °C/min.

It can be seen that the decomposition temperatures varied with the oxygen concentration in the atmosphere. Increase in oxygen partial pressure makes it more difficult for oxygen to escape and hinders sample decompose. Two-step decomposition indicates the decomposition of $Ca_{1.24}Co_{1.62}O_{3.86}$ to Co_3O_4, CaO and O_2 followed by the decomposition of Co_3O_4 to CoO and O_2.

Thermoelectric Properties

Figures 9 and 10 show the thermoelectric properties of $Ca_{1.24}Co_{1.62}O_{3.86}$ prepared by SHS measured at Oak Ridge National Laboratory. Figure 9 is a comparison of the Seebeck coefficients of $Ca_{1.24}Co_{1.62}O_{3.86}$ prepared by SHS and reported data of $Ca_{1.24}Co_{1.62}O_{3.86}$ prepared by other processes. From Figure 9, it can be seen that the Seebeck coefficient of powders from this work is much higher than those produced by other processes[10,11], especially at temperatures less than 300 °C.

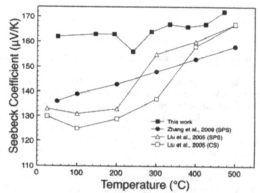

Figure 9. Seebeck coefficients of $Ca_{1.24}Co_{1.62}O_{3.86}$ prepared by different processes.

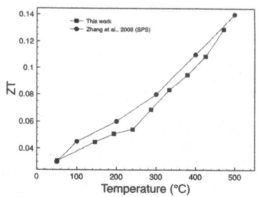

Figure 10. Figures of merit of $Ca_{1.24}Co_{1.62}O_{3.86}$ prepared by SHS and spark plasma sintering.

Electrical conductivity of our powder (~ 70 S/cm) is substantially higher than those prepared by convectional high temperature sintering (about 30 S/cm) but lower than those prepared by spark plasma sintering (~ 100 S/cm). Similar pattern was observed for power factor (~ 175 μW/mK^2). Due to the high Seebeck coefficient, sample prepared by SHS had figure of merit values, ZT close to the those prepared by spark plasma sintering method (Fig 10).

Mathematical Modeling

Temperature profiles calculated from our model at 30, 50, and 60 seconds are shown in Figure 11. Pellet is ignited from the left side of the pellet by an initial applied heat flux, which simulates graphite igniter. After ignition reaction front propagates evenly trough the pellet in both x and y-direction until it reaches the pellet surface. Reaction then advances in x-direction through the rest of the pellet. Maximum temperature reached by the model was 1,500 °C, which is about 200 °C higher than the measured value in experiments.

Figure 11. Temperature profile at 30, 50, and 60s after ignition.

Low thermal conductivities (0.45 W/(m K) for reactants and 0.28 W/ (m K)) results in a very narrow reaction front. Thermal conductivity was found to be the variable that affects the temperature profile of SHS reaction most. It governs the reaction propagation speed, reaction front thickness, and pellet cooling after reaction, among other factors. For example just a 15% increase of thermal conductivity of reactants resulted in reaction not propagating. This suggests that porosity of a pellet (thermal conductivity is strongly dependant on pellet porosity) affects SHS reactions.

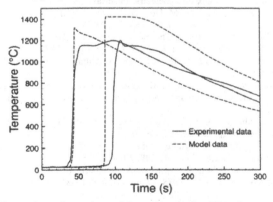

Figure 12. Comparison of experimental data with calculated data from our model.

Reaction front propagation speed agrees fairly well with experimental measurements (Fig 12). Currently our model results approximately 15% faster propagation speed. Also, it can be see that heating rate measured during experiments is consistent with model results. However, temperature reached about 200 °C higher in the model. This can be attributing to cracking of pellet during the SHS, which could result in faster cooling. In addition, evaporation of water due to decomposition of calcium hydroxide is not considered in the model. The evaporation of water is highly endothermic process and could play important role in governing temperature of SHS of calcium-cobalt oxide. Further investigation will be conducted to understand the reaction network.

CONCLUSIONS

We were able to produce highly pure $Ca_{1.24}Co_{1.62}O_{3.86}$ by self-propagating high-temperature synthesis followed by only 15 min post treatment. The decomposition temperature of prepared $Ca_{1.24}Co_{1.62}O_{3.86}$ increases with oxygen partial pressure in the atmosphere. Prepared sample had Seebeck coefficient higher than $Ca_{1.24}Co_{1.62}O_{3.8}$ prepared by other processes. Figure of merit, ZT was comparable to samples prepared by spark plasma sintering. Model of SHS reaction suggests that thermal conductivity (and porosity) play important role during SHS reaction.

ACKNOWLEDGEMENT

This work was partially supported by Lamar University Research Enhancement Grants under project No. CM1006 and Texas Hazardous Waste Research Center (THWRC) under contract No. 068LUB0969.

REFERENCES

[1]Fairbanks, J., Thermoelectric Developments for Vehicular Applications, U.S. Department of Energy: Energy Efficiency and Renewable Energy. Presented on: August 24, 2006.
[2]Annual Energy Review 2007, Report No. DOE/EIA-0384(2007), Energy Information Administration, http://www.eia.doe.gov/emeu/aer/pdf/pages/sec2.pdf.
[3]D. M. Rowe, CRC Handbook of Thermoelectrics, CRC Press LLC, Boca Raton, FL, 1995.
[4]I. Terasaki, Y. Sasago and K. Uchinokura, Large thermoelectric power in $NaCo_2O_4$ single crystals, *Phys. Rev. B*, **56**, R12685-R12687 (1997).
[5]G.J. Snyder, E.S. Toberer, Complex thermoelectric materials, *Nature Materials*, 7, 105-14 (2008).
[6]A. Masset, C. Michel, A. Maignan, and M. Hervieu, Misfit-layered cobaltite with an anisotropic giant magnetoresistance: $Ca_3Co_4O_9$, *Phys. Rev. B*, **62**, 166–75 (2000).
[7]H. Itahara and T. Tani, Highly-Textured Thermoelectric Oxide Polycrystals Synthesized by Reactive-Templated Grain Growth (RTGG) Method, *R&D Review of Toyota CRDL*, **39**, 63-70 (2004).
[8]Qiang Li, On the Thermoelectric Properties of Layered Cobaltates, *Mater. Res. Soc. Symp. Proc.*, **886**, F01-05.1- F01-05.12 (2006).
[9]M. Mikami, E. Guilmeau, R. Funahashi, K. Chong, D. Chateigner, Enhancement of electrical properties of the thermoelectric compound $Ca_3Co_4O_9$ through use of large- grained powder, *J. Mater. Res.*, **20**, 2491-97 (2005).
[10]Y. Liu, Y. Lin, Z. Shi, C. Nan, Z. Shen, Preparation of $Ca_3Co_4O_9$ and Improvement of its Thermoelectric Properties by Spark Plasma Sintering, *J. Am. Ceram. Soc.*, **88**, 1337-40 (2005).
[11]Y Zhang, J Zhang, Rapid reactive synthesis and sintering of textured $Ca_3Co_4O_9$ ceramics by spark plasma sintering, *J. Mater. Process. Technol.*, **208**, 70-74 (2008).
[12]B. B Kniha, B. Formanek, I. Solpan, Limits of Applicability of the "Diffusion-controlled Product Growth" Kinetic Approach to Modeling SHS, *Physica B*, **355**, 14-31, (2005).

EFFECT OF RARE-EARTH DOPING ON THERMOELECTRIC PROPERTIES OF POROUS SiC SYNTHESIZED BY SILICON CARBONIZATION TECHNIQUE

Yusuke Yamamoto, Hiroshi Mabuchi, Toshiyuki Matsui

Graduate School of Engineering, Osaka Prefecture University

Sakai, Osaka 599-8531, Japan

ABSTRACT

By means of the silicon carbonization technique using the graphite powder containing Dy_2O_3 or Gd_2O_3, porous SiC in which Dy or Gd were finely dispersed, was successfully synthesized. The magnetic impurity densities estimated from the analysis of the magnetic properties of the Dy/Gd doped samples. There is little difference between the effective carrier densities and the magnetic impurity densities for the Gd doped samples. In contrast, for the Dy doped samples, the effective carrier densities were much larger than the magnetic impurity densities. These results indicated that Dy might be mainly doped near the surface of the SiC wafer. Hence the effective carrier densities of the Dy doped samples in the near surface region were considered to significantly increase in contrast to the ND samples. As a result, the power factors of the Dy doped samples were much improved to exhibit the highest value of 5.8×10^{-5} W/mK^2 at 897 K.

INTRODUCTION

Materials having a considerably high Seebeck coefficient are expected to be used for thermoelectric devices which convert thermal energy into electric energy.[1] The efficiency of such devices, however, is not very high especially at elevated temperatures. Hence many researchers have still been looking for materials with excellent thermoelectric natures at high temperatures. Here, it is worthwhile mentioning that 3C-SiC is known to be a wide band gap semiconductor having good mechanical characteristics, heat resistance and corrosion resistance even at elevated temperatures.[2-4] Accordingly this material were expected to be one of the potential thermoelectric materials using at high temperatures. However SiC has also been known to have several disadvantages to be overcome for considering the practical usage for a thermoelectric system: the typical one of them is very high thermal conductivity. Here, it should be worthwhile mentioning that we have ever successfully synthesized the porous SiC with considerably low thermal conductivity by silicon carbonization technique. In spite of this, the physical properties of the porous SiC fabricated by using above mentioned process, still remain to be improved if the practical usage of the porous SiC for thermoelectric materials is considered.

It was reported that some of rare-earth compounds had desirable electronic structures for thermoelectric natures, that is, the 4f-energy levels of these compounds were located near the Fermi level.[1, 6] So it can be expected that the electronic structure of the porous SiC would be also modified by rare-earth element doping. However, the rare-earth elements are well known to be difficult to handle due to their severe oxidizability. In the present studies, we tried to synthesize the rare-earth doped porous SiC by the silicon carbonization technique. Silicon carbonization technique is one of the processes which enables to make the porous SiC from pure Si on near net-shaping. During the process, CO and CO_2 gases are generated in the furnace from the graphite and the residual oxygen under argon atmosphere. Then the following two reactions occurred:[5]

$$Si + 2CO = SiC + CO_2 \qquad (1)$$
$$Si + CO_2 = SiO(\uparrow) + CO \qquad (2)$$

The formula (1) shows the formation reaction of SiC by the CO gas, and the formula (2) shows the oxidation reaction of Si by CO_2 gas, causing the formation of pores within the starting material of Si wafer. Under this process, the SiC formation reactions proceed under the reduction atmosphere in the presence of CO gas. If rare-earth oxide compound can be deoxidized during the reaction process, the formation reaction of SiC and the doping reaction of rare-earth elements may be simultaneously occurred.

In the course of the present study, we tried to dope rare-earth elements for the porous SiC to improve the thermoelectric properties. In this paper, we will discuss the possibility of the Dy or Gd doping for the porous SiC, and the effects of their doping on the physical properties of the samples.

EXPERIMENTAL

The starting material of Si wafers (N-type, resistivity 0.02 Ωcm) was put into graphite cases filled with mixed powder of graphite (purity 99.98 %) and 0-40 wt% Dy_2O_3 (purity 99.5 %) or Gd_2O_3 (purity 99.9 %). In the present paper, the samples fabricated with only graphite powder are named non-doped (ND), and as well the samples using the mixed powder contained x wt% Dy_2O_3 powder are typically named Dyx. The graphite cases were heated up to 1663 K in vacuum, and kept at this temperature for 48 hours under around 101.3 kPa of the purged argon gas to induce the silicon carbonization reaction. The structural characterizations of the reacted samples were performed by X-ray diffraction (XRD). The microstructures of the samples (surface and cross section) were observed by a field emission scanning electron microscope (FE-SEM). The densities of the samples were measured by an Archimedes method. The magnetization behavior was characterized by a superconducting quantum interference device (SQUID) to confirm the Dy/Gd doping. Seebeck coefficients and electric conductivities were evaluated by a home made apparatus. The effective carrier densities determined by Hall effect measurement were also examined.

RESULTS

Structural Analyses

Figure 1 shows the XRD scans for the samples after the silicon carbonization reaction. The main phase for almost all the samples was identified to be 3C-SiC with a small amount of the second phase of Si or C. However, the weak diffraction peaks of unknown phase were observed in the Gd40 samples. The positions of the diffraction peaks of the 3C-SiC phase did not shifted in comparison with the calculated values.

The cross sectional microstructures of the ND, Dy40, Gd40 samples are typically shown Figs. 2(a-c). It was found that the porous structure was entirely formed in the samples. As clearly shown in the images the pore sizes of the Dy40 and Gd40 samples were bigger than those of the ND samples.

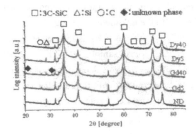

Fig. 1   X-ray diffraction scans for each sample.

Fig. 2   Cross sectional FE-SEM images of ND (a), Dy40 (b), Gd40 (c).

According to the density measurement, the densities and the apparent porosity of all the samples were unchanged, that is, around 1.7 g/cm³ 48 %, respectively. This indicates that the large pore and SiC net were formed in the samples fabricated by using mixed powder contained Dy_2O_3 or Gd_2O_3.

Magnetic Analyses

To confirm the Dy/Gd doping in the samples, the magnetic properties were measured, since Dy and Gd were known to be magnetic elements. Figures 3(a) and (b) show the M-H curves of each sample measured at 300 K and 5 K respectively. The magnetizations at 300 K of the ND and Dy/Gd doped samples, except for the Gd40 samples, decreased linearly with increasing the magnetic field. In addition, small spontaneous magnetization can be seen as well in the loops. These facts indicate that the samples exhibit diamagnetic behavior, but contained a small amount of ferromagnetic component. It should be noted here, the Gd40 samples had a larger amount of the ferromagnetic component than other samples. This may be due to the unknown magnetic phase, which was observed in the XRD scan for the Gd40 samples. The ND samples at 5 K also indicated diamagnetic behavior. In contrast, the magnetizations of all the Dy/Gd doped samples at 5 K gradually increased with increasing the magnetic field. This is considered to be typical paramagnetic behavior which is observed under the strong magnetic field at low temperature. This implies that the Dy or Gd atoms were successfully dispersed in the porous SiC without making clusters nor magnetic compounds.

Thermoelectric Analyses

Figure 4(a) shows the temperature dependence of the Seebeck coefficient of the ND and Dy/Gd doped samples. As is obvious in the negative sign of the Seebeck coefficient, all the samples exhibited N-type conduction behavior. The value of Seebeck coefficients of all the samples increased with increasing the temperature. In comparison with the Seebeck coefficients of the ND samples, those of the Gd doped samples were slightly small. On the other hand, those of the Dy doped samples were notably decreased with increasing the initial amount of the oxide powder.

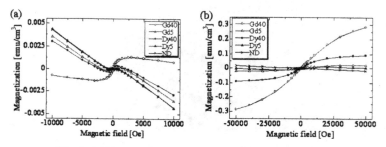

Fig. 3   M-H curves of each sample at 300 K (a) and 5 K (b).

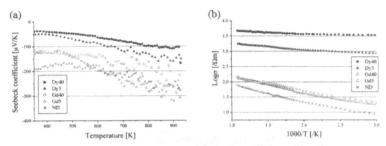

Fig. 4   Temperature dependence of Seebeck coefficients (a) and
temperature dependence of electric conductivities (b).

The temperature dependence of the electric conductivities is also shown in Fig. 4(b). The electric conductivities increased with increasing the temperature. The Gd doped samples exhibited the slightly larger values than the ND samples. However the conductivities of the Dy doped samples were about two order higher than those of the ND samples.

The change in the electric properties by the Dy/Gd doping, such as the decrease in the Seebeck coefficient and the increase in the electric conductivity, may suggest that the effective carrier densities of each sample were significantly changed.[1] Therefore, we evaluated the effective carrier density by hall effect measurement. Figure 5 shows the change in the effective carrier density in accordance with the initial amount of the oxide powder. Although the effective carrier densities of the Gd doped samples scarcely increased, those of the Dy doped samples remarkably increased with a small amount addition of the oxide powder. However, those values for both the samples almost remained constant even if the amount of the oxide powder increased. This is consistent with the change in the electric conductivities of each sample observed. Accordingly the estimated carrier nobilities of the samples were considered to be almost unchanged. These results suggested that the change of the electric conductivities of the Dy/Gd doped samples were caused by the change of the effective carrier

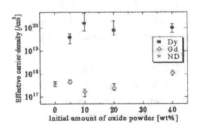

Fig. 5   Amount of oxide powder dependence of the effective carrier densities.

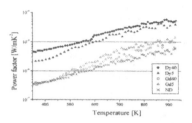

Fig. 6   Temperature dependence of the power factors.

densities.

Temperature dependences of the power factor of the samples were shown in Fig. 6. The power factors increased with increasing the temperature for all the samples. The power factors of the Gd doped samples exhibited the slightly larger values than those of the ND samples. The power factors of the Dy doped samples reached the highest value of 5.8×10^{-5} W/mK^2 at 897 K, which was almost ten times larger than those of the ND samples.

DISCUSSION

In accordance with the magnetic analysis, the ferromagnetic components were observed at 300 K for all the samples. These were observed even for the ND samples. Hence, this may be caused by the impurity elements in the graphite power. Since graphite powder was known to contain some impurities such as Fe, the carbonized SiC sample may also contain these magnetic impurities which resulted in exhibiting the ferromagnetic behavior. However the effect on the physical properties caused by this may be negligible, since the doping amount is much less than those of Gd or Dy.

The obvious paramagnetic behaviors were observed at 5 K for the Dy/Gd doped samples. Hereby, the total doping amount can be estimated by a Brilloiun fitting. In order to do this analysis, the diamagnetic component of the samples must be subtracted. The subtractions were carried out through the estimation from the gradient of the M-H curve of the ND samples. Since Dy or Gd usually exist as cations of Dy^{3+} or Gd^{3+}, the value of J for Dy and Gd were assumed to be 15/2 and 7/2 respectively. In addition, we assumed that the ferromagnetic component is much smaller than paramagnetic component at 5 K, because the magnetizations at 5 K are more than ten times larger than those of at 300 K. Figure 7(a) shows the fitting result as well as the measured magnetization data. The fitting curve of the Gd40 samples contained relatively large displacement which might be due to the presence of the ferromagnetic compound. For other samples, the calculated magnetization curves were basically well fitted to the measured ones. So the magnetic impurity densities of each sample could be estimated, which was shown in Fig. 7 (b). The magnetic impurity densities of the Dy/Gd doped samples except for the Gd40 samples were in the range of 10^{17} to 10^{18} cm^3. For the Gd doped samples, the magnetic

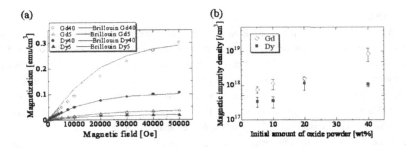

Fig. 7   Brillouin fitting results (a) and the relationship between the amount of the
oxide powder, and the estimated magnetic impurity densities (b).

impurity densities and the effective carrier densities were almost same, apart from the Gd40 samples
containing the unknown magnetic phase.

In contrast to this, the magnetic impurity densities of the Dy doped samples were about two
orders lower than the effective carrier densities of the Dy doped samples. These discrepant results may
be explained by difference in the regions in which the magnetic properties measured and the hall effect
measured. That is, the data of the magnetic properties were observed from the whole region of the SiC
wafer. In contrast, the result of the hall measurements must be dominantly affected by the surface
regions of the SiC wafer, if the SiC wafers were inhomogeneous. Assuming that Dy was mainly doped
only near the surface regions of the SiC wafer, the effective carrier density which was determined by
Van der Pauw measurements was expected to be larger than the impurity density that was estimated by
the magnetic measurement. Accordingly, it can be speculated that Dy was mainly doped only near the
surface region. In other words, the actual density of the doped Dy near the surface region is much
larger than those of the estimated magnetic measurements. These results indicate that the effective
carrier of the Dy doped samples of near the surface is drastically increased in contrast to the ND
samples. As the result, the power factors of the Dy doped samples were drastically increased in
contrast to the ND and Gd doped samples.

Some energy levels of the doped Dy/Gd possibly exist near the conduction band because these
samples have the very high effective carrier densities and N-type conduction behavior. Therefore the
increment of power factors observed were mainly caused by the effective carrier density modified, not
having desirable electronic structure. The thermoelectric properties of the porous SiC may be improved
by the controlling Fermi level, which may have the desirable electronic structure.

CONCLUSION

By means of the silicon carbonization technique using the graphite powder containing Dy_2O_3 or Gd_2O_3, the porous SiC in which Dy or Gd were finely dispersed was successfully obtained. The magnetic impurity densities of the Dy/Gd doped samples were estimated by Brillouin fitting of the magnetization curves at 5 K. The estimated magnetic impurity densities of the Gd doped samples were almost same as the effective carrier densities determined by a hall measurement. In contrast, for the Dy doped samples the magnetic impurity densities of the Dy doped samples were about two orders lower than the effective carrier densities. This result indicated that Dy might be mainly doped near the surface of the SiC wafer, which the effective carrier of the Dy doped samples of near the surface is drastically increased in contrast to the ND samples. Hereby, the Dy doped samples were thought to have more suitable amount of the effective carrier densities as thermoelectric materials. As a result, the power factors of the Dy doped samples reached the highest value of 5.8×10^{-5} W/mK^2 at 897 K, which was almost ten times larger than those of the ND samples.

REFERENCES

[1] G. Mahan, B. Sales, J. Sharp, Thermoelectric Materials New Approaches to an Old Problem, *Physics Today* **50** (3), pp. 42-47 (1997)

[2] A. I. Shelykh, B. I. Smirnov, T. S. Orlova, I. A. Smirnov, A. R. de Arellano-Lopez, J. Martinez-Fernandez, F. M. Varela-Feria, *Physics of the Solid State* **48** (2), pp. 229-232 2006

[3] M. Uehara, R. Shiraishi, A. Nogami, N. Enomoto, J. Hojo, SiC–B$_4$C composites for synergistic enhancement of thermoelectric property, *Journal of the European Ceramic Society* **24** (2), pp. 409-412 (2004)

[4] K. Koumoto, M. Shimohigoshi, S. Takeda, H. Yanagida, Thermoelectric energy conversion by porous SiC ceramics, *Journal of Materials Science Letters* **6** (12), pp. 1453-1455 (1987)

[5] M. Masuda, H. Mabuchi, H. Tsuda, T. Matsui, K. Morii, Thermoelectric Properties of 3C-SiC Produced by Silicon Carbonization, *Materials Science Forum* **389-393** (1), pp. 763-766 (2002)

[6] G. D. Mahan, J. O. Sofo, The best thermoelectric, *Proceedings Of The National Academy Of Sciences Of The United States Of America* **93**, 7436(1996)

POWDER SYNTHESIS, CHARACTERIZATION AND SINTERING BEHAVIOR OF LITHIUM TITANATE

Srinivasan Ramanathan
Material Processing Division
Bhabha Atomic Research Centre
Mumbai - 4000 85
Maharastra
India

ABSTRACT

A comparative study of the sintering behavior of the lithium titanate ($Li_2 TiO_3$) powders formed by combustion of the nitrate (lithium and titanium)-glycine gel (powder-A) and reaction of the mixture of lithium carbonate with titania (powder-B) was carried out. The calcination temperature to form chemically pure and nano-crystalline powders as obtained by DTA- TG and XRD was found to be 600^0C for powder -A and 750^0C for powder-B. The dry ground powders contained coarse agglomerates and were wet ground under an optimized dispersion condition (pH ~9, solid content 20wt.%). The agglomerate size of the ground slurries were ~0.6µm for powder -A and ~2.6µm for powder-B. A study of the shrinkage behavior by dilatometry exhibited better sinter-ability for powder –A. A detailed study of the pore size distribution in the green compacts by mercury intrusion porosimetry revealed that the compacts from powder-A exhibit finer size distribution (average size ~0.1µm) than the compacts from powder-B ((average size ~0.4µm) even though their green compact densities were in the range of 50 to 55%T.D. After sintering at a temperature of 1100^0C for 2hrs, compacts of powder-A and powder-B exhibited a sintered density of 92 and 78% T.D. respectively. The difference in sintering behavior of the two powders was correlated to the difference in the pore size distribution in the green compact and its evolution during heating.

Key words: sinter/sintering, combustion synthesis, pores/porosity

INTRODUCTION

Lithium compounds such as zirconate, aluminate, silicate and titanates of lithium find application as potential solid breeders for tritium in fusion reactor.[1,2] Among these systems, fine grained (3-5µm) and porous (~20%) lithium titanate (Li_2TiO_3) has been reported to be the ideal material for this application due to its good tritium recovery at low temperature, chemical stability, low tritium inventory etc.[3-7]. The ceramics is required in the form of pebbles and are processed through the standard ceramic processing route using powder. There are various methods for the synthesis of powders such as solid state reaction between lithium oxide /carbonate and titanium dioxide, soft chemical routes such as sol - gel process (using inorganic or organic or hybrid precursors containing lithium and titanium ions followed by hydrolysis) and gel combustion (using nitrates of lithium, titanium and organic compounds such as glycine or citric acid or urea).[8-18] A study of the sintering behavior of the powders is essential in tailoring their characteristics to form sintered bodies with desired specifications. Choong-Hwan Jung has studied the sintering behavior of the powders formed by the reaction of lithium carbonate-titania and combustion of the gel of the metal nitrate – glycine using dilatometry and found the gel combustion route powder sinters at a lower temperature.[17] Better sinter-ability is generally expected when the distance over which mass transport has to occur is less, which in turn is decided by the pore size distribution of the green compact. The agglomerate size distribution and specific surface area of the powder are the two important parameters that influence the pore size distribution of green compact. No study has been yet reported correlating the agglomerate size of the particles used and pore size distribution of the green compacts formed with sinter-ability of the lithium titanate powders formed by the above mentioned processes (i.e., gel combustion and solid state reaction) and hence has been studied. The conditions for formation of chemically pure nano-crystalline powder by the above two techniques has been studied using TG-DTA and XRD. The powder formed generally are soft agglomerated and need to be de-agglomerated by wet milling under optimized dispersion conditions which is obtained from zeta potential and rheological studies. The agglomerate and pore size distribution have been characterized using laser light scattering technique and mercury intrusion porosimetry. The pore size evolution during sintering has been studied and the sinter-ability has been correlated with the pore structure.

EXPERIMENTAL

Required amount of lithium nitrate, titanyl nitrate and glycine were taken from stock solutions of known concentration (corresponding to a batch size of 100g of the oxide compound) and were concentrated on a laboratory heater to form into a viscous gel that dried upon prolonged heating at 80^0C in an air oven. The stoichimetric amount of glycine fuel required for combustion was calculated using the following reaction:

$$9\ MNO_3 + 5\ NH_2CH_2COOH \rightarrow 1.5\ M_2O + 10\ CO_2 + 12.5\ H_2O + 4\ N_2$$

Also required amount of lithium carbonate and titania (corresponding to a batch size of 100g of the oxide compound) were wet-mixed/milled for homogenization in a planetary mill for 2 hours and subsequently dried at 80^0C to form in to a cake. Both the oven dried gel and cake were used for evaluation of their thermal decomposition behavior by TG-DTA and phase evolution by XRD. Based on the TG-DTA and XRD results the temperature of decomposition of the gel and cake was fixed. The powders were prepared by both combustion of the nitrate - glycine gel (Powder -A) on a heater at about at 300^0C and solid state reaction of lithium carbonate - titania (Powder -B) at about at 750^0C. The powders formed were de-agglomerated by dry and wet grinding under identical conditions (pH 9-10 with 20 weight percent solid content for 2 hour). The dispersion condition for aqueous - wet grinding of the powders formed was obtained by studying the zeta potential of the aqueous suspensions with varying pH using laser Doppler velocimetry. The viscosity variation with shear rate was studied using a cone and plate viscometer. The wet grinding was carried out in a planetary mill at 200 rpm using alumina pot (250cc) and balls (50 balls of 10mm diameter). The agglomerate size of the ground powders was determined using laser light scattering technique while densification behavior was studied by measuring the density of the sintered compacts by the Archimedes method. The shrinkage behavior of the compacts from both powders was studied using a dilatometer with a rate of heating of 4^0C per minute in air. The pore size distribution in green and sintered compacts was studied using mercury intrusion porosimetry.

RESULTS AND DISCUSSION

The DTA-TG patterns for the nitrate-glycine gel and the carbonate-oxide mixture are shown in Fig.1(a and b). It is obvious that the gel exhibits 80 percent loss in weight accompanied by a sharp and intense exotherm around a temperature of about 230^0C which is attributed to combustion reaction. The gel was amorphous while the product after combustion was nano-crystalline Li_2TiO_3 phase (pc-pdf no.33-0831) (Fig.2). There was a little loss in weight (5wt.%) in the temperature range of 250 to 600^0C indicating the loss of residual volatiles left behind due to short time of combustion reaction and hence the powder was calcined at 600^0C for 1 hour. The reaction of carbonate-oxide mixture took place over a broad temperature range of 400 to 750^0C accompanied by an endotherm and corresponding loss in weight of about 27% as expected from the reaction:

$$Li_2CO_3 + TiO_2 \dashrightarrow Li_2TiO_3 + CO_2\uparrow$$

Both the heat effect and loss in weight in the above reaction are much less compared to the combustion reaction of the gel. The product formed was found to be phase pure Li_2TiO_3 (Fig2)

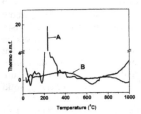

Fig.1.A. DTA plots of the precursors for powders A and B

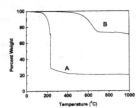

Fig.1.B. TG plots of the precursors for powders A and B

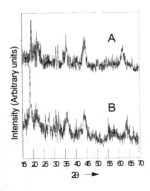

Fig.2. XRD patterns for the powders

In case of gel combustion, the aqueous solution containing nitrates of lithium, titanium and glycine was dried at 80^0C in an air oven so that it could be powdered and used for thermal analysis investigation. The dried gel forms into a viscous liquid by continued heating and led to frothing followed by smooth burning with vigorous evolution of a large amount of gaseous product forming a porous net work of the solid compound. It was essential to use proper size of the reaction vessel with a lid of stainless steel wire mesh to avoid loss of the fine particles of the product by the evolved gases. The evolution of large amount of gaseous product resulted in formation of voluminous, porous and soft agglomerated nano-powder while the exothermicity of the reaction made it to go to completion without supply of additional heat. It is important to note that combustion reaction evolves only neutral gases to the environment. The compositional homogeneity expected to be present in the gel network in molecular level facilitates easy formation of compound. The as formed porous powder was calcined at 600^0C for 1 hour to form into a chemically pure oxide compound (powder-A). In case of lithium carbonate – titania reaction the process was found to be endothermic and occurs in the temperature range of 400 to 750^0C and hence the reaction was carried out at 750^0C for 3hrs. Generally, solid state reaction where mass transport occurs by diffusion, takes long time for the reaction to go to completion. However, in this case the reaction completed even in 1 hour as indicated by weight loss and XRD data. Even though Li_2CO_3 melts at a temperature of 680^0C, the reaction sets in around 400^0C onwards and the rate of reaction is maximum around 600^0C (Fig.1A & B). This could be attributed to the lowering of the melting point due to reaction with titania. Once melting occurs, the reaction is expected to be vigorous due to infiltration of the liquid in to the agglomerates of titania. Hence the reaction was carried out at 750^0C for 3 hrs to form into the oxide compound (powder-B).

The powders as formed were dry ground in a planetary ball mill till the powder mass forms a cake along the wall of the pot (took about 30 minutes for both the powders A & B). The particle size distribution of the dry ground powders is shown in Fig.3.

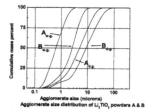

Fig.3. Agglomerate size distribution of dry and wet ground powders (A and B) of Li_2TiO_3

$A_{d.g.}, B_{d.g}$ -dry ground; $A_{w.g.}, B_{w.g.}$-wet ground)

It is obvious that the powder contains agglomerates with substantial fraction (>30%) with size in the range of 10 to 50μm which are detrimental for sintering into bodies with homogeneous grain structure. Hence they need to be eliminated only by wet grinding. The role of dispersion characteristics (i.e., zeta potential and viscosity) of the slurries on the effective wet grinding has been emphasized by the work of Hanoviet etal[19]. In slurries, the particles should be sufficiently charged so that they remain dispersed due to columbic repulsion. As it is a property of the surface interaction with the medium and that small quantities of surface adsorbents can drastically modify the numerical value, which depends upon the method of preparation. Hence a study of this property was carried out for both the powders A and B. The study on variation of zeta potential with pH of the suspensions of both the powders showed it to be maximum in the pH range of 9 to 10 (-30mV to -35mV) (Fig.4).

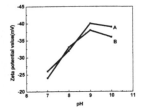

Fig.4. Zeta potential variation with pH for aqueous suspensions of powders A & B

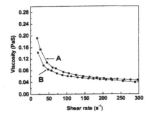

Fig.5. Viscosity variation with shear rate for aqueous slurries of powders A & B (20 wt%)

It is obvious that both the powders exhibit similar charging behavior even though the exact numerical values are slightly varying and the values found are in agreement with the reported literature.[14] Hence the pH for slurry formation for wet grinding was fixed as 9 to10 using ammonia. Even at the point of maximum zeta potential, flocculates are said to exist in concentrated slurries which is indicated by its flow behavior which in turn is strongly influenced by the solid loading. As the effective grinding require slurries with minimum pseudo plasticity or Newtonian flow characteristics (Fig.5) the solid loading used was only 20 weight percent for both the powders A & B. The viscosity was found to vary from 0.2 to 0.05 Pas in the shear rate range of 10 to 300cm^{-1} for the typical slurries used at a pH of 10 (Fig.5). It is seen that the slurry of powder B exhibited slightly lower value than that from powder A. The particle size distribution of the slurries after wet grinding for 2 hour (Fig.3) exhibited absence of agglomerates with size above 10μm. However, the powder-A exhibited finer size (D_{50} ~0.6μm) than powder-B (D_{50} ~ 2.6μm) exhibiting the fact that the agglomerates from the former route are softer compared to those from the later route.. The green density of the compacts were 52 and 55 percent of theoretical value.

The sintering behavior of the powders as evaluated by the dilatometry data exhibited on-set of sintering at a lower temperature for the powder -A (Fig.6) which is in agreement with the finding of Choong Hwan Jung etal.[16,17] To understand the reason for this behavior, a detailed investigation of the pore size evolution in the compacts during sintering was carried out using mercury intrusion porosimetry. The pore size distribution of the green compacts of both the powders A & B and the evolution during sintering is shown in Fig.7.

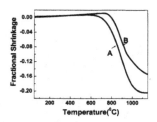

Fig.6. Dilatometric shrinkage data of compacts from powder –A & B

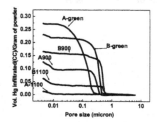

Fig.7. Pore size evolution in compacts of powder A and B during sintering

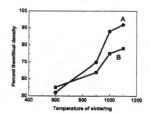

Fig.8. Sintered density variation with temperature for powders A & B

It is obvious that even though the green densities remained essentially the same, the compacts from powder-A exhibited an average pore size around 0.1μm while that from powder-B exhibited a higher value of 0.4μm which could be attributed to the difference in their agglomerate size and specific surface area. Upon sintering, due to the faster rate of elimination of finer pores, the average pore size of the remaining porosity increased for the compacts of powders from both the routes. Due to finer pore size, the extent of elimination of pores is more in compacts from powder-A than that from powder-B. Thus the compacts from powder-A sintered to a density of 92% T.D. at 1100°C while that from powder-B sintered to only 78%T.D. Thus it is obvious that the presence of a larger volume fraction of finer pores in powder-A contributed to its better sinter-ability which is the main finding of this study. The densification behavior plots as a function of temperature for the compacts sintered for 2 hrs at varying temperatures for both the powders exhibited slowing down of extent of densification with progress of sintering (Fig.8) which is also in agreement with the pore size evolution behavior. It is obvious, as expected, that the agglomerate size distribution of the powders play a major role in the pore size distribution of the green compacts.

CONCLUSIONS

Lithium titanate (Li_2TiO_3) powders were synthesized by combustion of the nitrate (lithium & titanium) -glycine gel and reaction of lithium carbonate-titania. The study on thermal decomposition and phase evolution behavior revealed that exothermic combustion sets in above 230^0C while the endothermic carbonate-oxide reaction sets in over a broad range of temperature (400- 750^0C). In both cases nano-crystalline oxide compounds formed. They were agglomerated and were dispersed by wet grinding using aqueous slurries at a pH of 10 and solid content of 20 weight percent. The results of the dilatometric investigation indicated better sinterability for the powder from gel combustion route (powder-A) over carbonate-oxide reaction route (powder-B). The better sinter-ability of the gel combustion powder is attributed to its finer agglomerate size and presence of a large fraction of porosity in the finer size range in the green microstructure of the compact, as brought out by the particle and pore size distribution studies.

ACKNOWLEDGEMENTS

The author thanks Mr. S. Kolley and M.R. Gonal for their help in TG-DTA and XRD investigations. The author thanks Dr. N. Krishnamurthy, Head, High temperature Materials Processing, Dr I.G. Sharma, Head Materials Processing Division and Dr A.K. Suri, Director Materials group, for their keen interest and encouragement for this work.

REFERENCES

1. C.E. Johnson, G.W. Hollenberg, N. Roux, H. Watanabe, Fusion Eng. Des. **8**, 145 (1989)
2. P.A. Finn, K. Kurasawa, S. Nasu, S. Noda, T. Takahashi, H. Takeshita, T. Tanifuji, H. Watanabe, in Proceedings of 9th IEEE Symposium on Engineering Problems of Fusion Research, vol.II, p.1200 (1981)
3. J.M. Miller, H.B. Hamilton, J.D. Sullivan, J. Nucl. Mater. **212-215**, 877 (1994)
4. N. Roux, J. Avon, A. Floreancig, J. Mougin, B. Rasneur, S. Ravel, J. Nucl. Mater. **233-237**, 1431 (1996)
5. R. Gierszewski, Fusion Eng. Des. **39 & 40**, 739 (1998)
6. T. Kawagoe, M. Nishikawa, A. Baba, S. Beloglazov, J. Nucl. Mater. **297**, 27 (2001)
7. K. Tsuchiya, H. Kawamura, T. Takayama, S. Kato, J. Nucl. Mater. **345**, 239 (2005)
8. C.E. Johnson, K.R. Kummerer, E. Roth, J. Nucl. Mater. **155-157**, 188 (1988)
9. M.A. Gulgun, M.H. Nguyen, W.M. Kriven, J. Am. Ceram. Soc. **82, 3**, 556 (1999)
10. M.H. Nguyen, S.J. Lee, W.M. Kriven, J. Mater. Res. **14, 8**, 3417-26 (1999)
11. S.J. Lee, W.M. Kriven, J. Am. Ceram. Soc. **81, 10**, 2605-12 (1998)
12. S.J. Lee, E.A. Benson, W.M. Kriven, J. Am. Ceram. Soc. **82, 8**, 2049 -55 (1999)
13. S.J. Lee, M.D. Biegalski, W.M. Kriven, J. Mater.Sci. **14, 7**, 3011-16 (1999)
14. S.J. Lee, J. Ceram. Process. Res. **9, 1**, 64-67 (2008)
15. C.H. Jung, J.Y. Park, J. Nucl. Mater. **341**, 148-52 (2005)
16. C.H. Jung, J.Y. Park, S.J. Oh, H.K Park, Y.S. Kim, D.K. Kim, , J.H. Kim, J. Nucl. Mater. **253**, 203-212 (1998)
17. C.H. Jung, J. Nucl. Mater. **341** 148-52 (2005)
18. C.ZH. Jung, J.Y. Park, W.J. Kim, W.S. Ryu, S.J. Lee, Fusion Eng. Des. **81**, 1039-44 (2006)
19. Hanoviet David Houivet, Jaffar El Fallah and Jean Marie Haussonne, J. Am. Ceram. Soc., **85, 2**, 321-28 (2002)

PROCESSING OF TITANIA NANOCERAMICS VIA CONVENTIONAL SINTERING, TWO-STEP SINTERING AND TWO-STEP SINTERING ASSISTED BY PHASE TRANSFORMATION

Zohreh Razavi Hesabi
Materials and Energy Research Center, P.O. Box 14155-4777, Tehran, Iran
Zohreh.rezavi@gmail.com

Mehdi Mazaheri
École Polytechnique Fédérale de Lausanne, Institute of Physics of the Condensed Matter, Station 3, 1015 Lausanne, Switzerland
Mehdi.mazaheri@epfl.ch and mmazaheri@gmail.com

ABSTRACT
In the present study, the novel technique of two-step sintering was applied on TiO_2 ceramic nanopowder to suppress accelerated grain growth during densification. A dense, uniform and ultrafine-grain ceramics were hence fabricated at a temperature lower than that of the conventional methods. While the grain size of is ~2 μm for full dense structures produced by conventional sintering, application of two-step sintering has led to a remarkable grain size decline to ~250 nm. Two-step sintering accompanied with anatase to rutile phase transformation led to formation of structure with an average grain size of ~100 nm.

INTRODUCTION

Synthesis and sintering of nanocrystalline ceramic powders have attracted much attention due to their promising properties[1,2]. The high active surface area of nanopowders results in lowering sintering temperature relative to coarser powders. Although low temperature sintering suppresses the grain growth, high density of interfaces and grain boundaries in nanocrystalline powders leads to accelerated grain growth during sintering. As a thermodynamical point of view, the lower the density of grain boundaries becomes, the lower the stored energy and the higher the structural stability would be[3]. Consequently a highly significant grain growth of nanopowders would be observed. Not only high interface area of nanopowders provides unique properties but also high density of grain boundaries of bulk nanomaterials brings them superior structural and functional properties[4]. To prohibit grain growth during densification there are different ways. For instance, Mazaheri et al.[5] used pressure assisted sintering to fabricate ultra-fine grained (UFG) ZnO structures. Shen et al.[6] used fast low temperature firing by spark plasma sintering to produce UFG alumina structures. While Tekeli et al.[7] added a second phase to hamper grain growth during sintering of nanocrystalline zirconia powder. By taking the benefits of drag forces by dispersoids, the grain size of 8 mole % yttria stabilized zirconia was significantly reduced. Although the aforementioned methods are successful to prohibit grain growth, sophisticated equipments and special conditions that have to be met limit using of them to produce UFG ceramic without deteriorating functional performance. For instance, Sakka et al.[8] showed that with the addition of titania and magnesia with an average particle size (APS) of 30 and 50 nm to 3 mol% yttria doped tetragonal zirconia (3YTZ) with APS of 70 nm, not only the grain size but also the ionic conductivity of the system decreased due to the suppression of oxygen ion mobility. Tekeli et al.[9] showed that with increasing the amount of Al_2O_3 particles (> 1 wt%) to produce finer cubic zirconia, the sinterability was degraded. Additionally Mori et al.[10] demonstrated that with the increasing the amount of Al_2O_3 particles the ionic conductivity of 8YSZ at 1000 °C decreased. Formation of insulating phases can cause degradation of ionic conductivity[11].

To avoid deteriorating effect of the second phase and dopants on sintering behavior of nanopowders and properties of sintered pieces, one can use the novel technique of two-step sintering to hinder grain growth that usually occurs at the final firing stage[12]. In this method without the addition of any dispersoids by using drag forces of triple junctions as obstacles against grain boundary migration, dense UFG ceramics could be produced. This method modifies the sintering regimes by high temperature firing followed by rapid cooling to lower temperature and long term soaking. For instance, Mazaheri et al.[4] used TSS technique to control the grain growth of 8YSZ nanopowder during sintering. Using this method, the grain size was decreased from ~2.15 μm to ~300 nm. As a consequence, up to ~ 100 % increase in the fracture toughness was observed (i.e. from 1.61 to 3.16 MPa.m$^{1/2}$).

To succeed in two-step sintering, a critical density should be obtained during the first step. Under this condition, residual pores become subcritical and unstable against shrinkage. These pores can be filled as long as grain boundary diffusion allows. The lower the temperature of first and second stage is, the finer the microstructure could be achieved. For instance, authors showed elsewhere[13] that by employing cold isostatic pressing for fabrication of green bodies, the sintering temperatures in TSS regimes of Al_2O_3 nanopowder were decreased to lower temperatures resulting in finer microstructure. It was attributed to smaller average pore size and narrower size distribution through green bodies. The question arises as to whether sintering temperatures can be decreased any further to obtain finer microstructure. Interestingly, Kumar et al.[14] densified nanocrystalline TiO_2 at lower temperatures by assisting anatase to rutile phase transformation. The increased mobility of the atoms during the phase transformation enhances the sintering rate at lower temperatures in which fine structure with near theoretical density could be produced. So it seems that by taking higher mobility of atoms near transformation temperature finer structure could be obtained in systems exhibiting phase transformation.

However, there is no systematic investigation on suppression of grain growth based on simultaneous two-step sintering and phase transformation assisted sintering. In the present research study, different TSS regimes were conducted to suppress grain growth in TiO_2 system showing anatase-to-rutile phase transformation.

EXPERIMENTAL PROCEDURE

TiO_2 nanopowder (P25, Degussa Co., Frankfurt, Germany) with the mean particle size of 19 nm was uniaxially pressed under 100 MPa and pellets with average green density of ~0.53 of theoretical one were obtained. Conventional sintering (CS) was conducted between 500-1000°C for 1 h in air with a heating ramp of 5°C min^{-1}. For two-step sintering (TSS) at first samples were sintered at 800 and 750 °C with for 1 h, then cooled to 700°C and held up to 25 h. To determine the grain size of sintered pellets, fracture surface of samples were studied using SEM (Philips XL30, Netherlands).

To determine the amount of anatase to rutile phase transformation during sintering, XRD analysis was performed in the 2θ range from 24° to 28° to distinguish anatase (101) occurring at 2θ=27.5° from rutile (110) occurring at 2θ=25.4° phase.

RESULTS AND DISCUSSION

Fig. 1 shows the effect of sintering temperature on relative density and grain size of TiO_2 compacts sintered conventionally. As the sintering temperature increased from 800 to 1000°C, an abrupt increase in grain size was observed while the density just increased from ~91% to ~98%. It indicates the significant grain growth at the final stage of sintering. Phase analysis of sintered samples showed the occurrence of anatase to rutile transformation during sintering. As shown, with increasing temperature the amount of rutile content increased. At the final stage of sintering accompanied with significant grain growth, phase transformation was completed and rutile content increased from ~88% to ~100%.

To suppress the grain growth, two-step sintering regime was employed. Fig. 2 shows changes of density and grain size as a function of holding time for TSS regime conducted at T1=800 °C for 1 h followed by cooling and holding at T2=700 °C. As shown, while densification continued no significant grain growth was occurred. As a consequence, nearly full-dense sample with grain size of 250 nm was obtained. In comparison with conventional sintering grain size was decreased from 2 µm to 250 nm. Negligible grain growth at the final stage of sintering could be related to the pinning of grain boundaries by triple junctions[13].

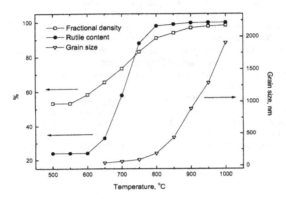

Fig. 1. Effect of sintering temperature on density and grain size of TiO_2 nanopowder compacts accompanied with anatase to rutile phase transformation.

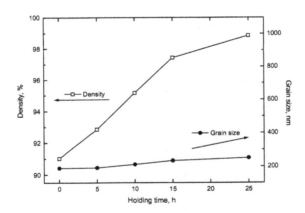

Fig. 2. Density and grain size as a function of holding time at 700 °C for samples firstly sintered at T1=800 °C for 1 h.

Referring to Fig. 1, rutile content at the starting point of second sintering step is ~98% (samples were firstly sintered at 800 °C for 1 h). Accordingly, second step of two-step sintering was conducted without interfering of phase transformation. Results obtained in the preset study (grain size and rutile content) are similar to those reported by lee at al.[14] for samples sintered by spark plasma sintering. They also obtained a full dense structure with the average grain size of ~200 nm, using the same TiO_2 nanopowder. It shows that TSS in terms of grain growth suppression is comparable with expensive SPS technique.

To take the benefits of phase transformation at the final stage of sintering, first step temperature was decreased to 750 °C. Therefore, the amount of rutile content at the starting point of second step in TSS regime was ~ 88%. During holding at the second step phase transformation was completed while negligible grain growth was observed. Successfully, the grain size of samples sintered via TSS assisted by anatase to rutile phase transformation was decreased to ~100 nm which is 5 times less than that obtained by previous TSS regime.

Fig. 3 shows the final microstructure of dense samples under TSS conditions in comparison with conventionally sintered one. As shown, a significant decrease in grain size was obtained. TSS accompanied with phase transformation resulted in formation of finer microstructure with an average grain size of 100 nm. Fig. 4 shows effect of phase transformation on densification and structural evolution of TiO_2 nanopowder sintered differently. Transformation of anatase to rutile during two-step sintering

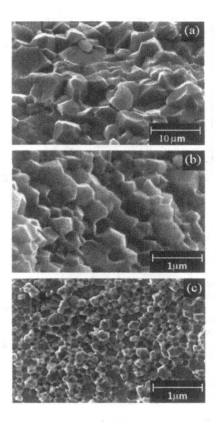

Fig. 3 SEM micrographs of the samples sintered (a) at 1000°C for 1 h, (b) at T1 = 800 °C for 1 h and T2 = 700 °C for 30 h and (c) at T1 = 750 °C for 1h and T2 = 700 °C for 30 h.

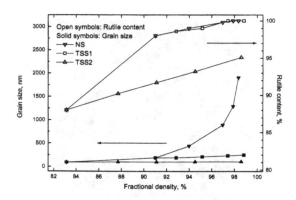

Figure 5. Effect of phase transformation (anatse to rutile) on densification and grain growth of compacted TiO$_2$ nanopowder.

CONCLUSION

In summary, the present study shows the formation of nanograins during two-step sintering of titania nanoceramics assisted by anatase-to-rutile phase transformation. The grain sizes of conventionally sintered samples were ~2 μm. With employing two-step sintering a remarkable decrease in the grain sizes down to ~250 nm was obtained. Enhancement of atomic mobility near phase transformation temperature made densification continued during TSS at lower temperatures. By TSS assisted phase transformation, a structure with finer grain sizes of around 100 nm was obtainable.

REFERENCES

[1]M.T. Swihart, Vapor-phase Synthesis of Nanoparticles, *Current Opinion in Colloid and Interface Science*, **8**, 127-133(2003).

[2] V.V. Srdic, M. Winterer, and H. Hahn, Sintering Behavior of Nanocrystalline Zirconia Doped with Alumina Prepared by Chemical Vapor Synthesis, *J. Am. Ceram. Soc.*, **83**, 1853-60 (2000).

[3]J. H. Driver, Stability of Nanostructured Metals and Alloys, *Scripta Materialia*, **51**, 819-823 (2004).

[4]M. Mazaheri, M. Valefi, Z. Razavi Hesabi, and S.K. Sadrnezhaad, Two-step Sintering of Nanocrystalline 8Y$_2$O$_3$ stabilized ZrO$_2$ Synthesized by Glycine Nitrate Process, *Ceram. Int.*, **35**, 13-20 (2009).

[5]Mehdi Mazaheri, S.A. Hassanzadeh-Tabrizi, and S.K. Sadrnezhaad, Hot Pressing of Nanocrystalline Zinc Oxide Compacts: Densification and Grain Growth During Sintering, Ceram. Int., **35** 991-995 (2009).

[6]Z. Shen, M. Johnsson, Z. Zhao, and M. Nygren, Spark Plasma Sintering of Alumina, *J. Am. Ceram. Soc.*, **85**, 1921-27 (2002).

[7]S. Tekeli, M. Erdogan, and B. Aktas, Microstructural Evolution in 8 mol% Y$_2$O$_3$-Stabilized Cubic Zirconia (8YSCZ) with SiO$_2$ Addition, *Mater. Sci. Eng. A***386**, 1-9 (2004).

[8]Y. Sakkaa, T.S. Suzukia, T. Matsumotob, K. Moritaa, K. Hiragaa, and Y. Moriyoshib, Effect of Titania and Magnesia Addition to 3 mol% Yttria Doped Tetragonal Zirconia on Some Diffusion Related Phenomena, *Solid State Ionics*, **172**, 499-503 (2004).

[9]S. Tekeli, Fracture Toughness (KIC), Hardness, Sintering and Grain Growth Behavior of 8YSCZ/Al$_2$O$_3$ Composites Produced by Colloidal Processing, *J. Alloys Comps*, **391**, 217–224 (2005).

[10]M. Mori, M. Yoshikawa, H. Itoh, and T. Abe, Effect of Alumina on Sintering Behavior and Electrical Conductivity of High-Purity Yttria-Stabilized Zirconia, *J. Am. Ceram. Soc.*, **77**, 2217-2219 (1994).

[11]T. Chen, S. Tekeli, R.P. Dillon, and M.L. Mecartney, Phase Stability, Microstructural Evolution and Room Temperature Mechanical Properties of TiO$_2$ Doped 8 mol% Y$_2$O$_3$ Stabilized ZrO$_2$ (8Y-CSZ), *Ceram. Int.*, **34**, 365-370 (2006).

[12]I.W. Chen and X.H. Wang, Sintering Dense Nanocrystalline Oxide with-out Final Stage Grain Growth, *Nature*, **404**, 168-71 (2000).

[13]Z. Razavi Hesabi, M. Haghighatzadeh, Mehdi Mazaheri, Dusan Galusek, and S.K. Sadrnezhaad, Suppression of Grain Growth in Sub-micrometer Alumina via Two-step Sintering Method, *J. Europ. Ceram. Soc.*, **29**, 1371-1377 (2009).

[14]K. -N. P. Kumar, K. Keizer, A. J. Burggraaf, T. Okubo, H. Nagamoto, and S. Morooka, Densification of Nanostructured Titania Assisted by a Phase Transformation, *Nature*, **358**, 48-51 (1992).

[15]Y.I. Lee, J.-H. Lee, S.–H. Hong, and D.–Y. Kim, Preparation of Nanostructured TiO$_2$ Ceramics by Spark Plasma Sintering, *Materials Research Bulletin*, **38**, 925-930 (2003).

STRENGTH OF N- AND P-TYPE SKUTTERUDITES[1]

A. A. Wereszczak,* M. E. Ragan,* K. T. Strong, Jr.,* P. J. Ritt,** H. Wang,**
J. R. Salvador,*** and J. Yang****

*   Ceramic Science and Technology
**   Diffraction and Thermophysical Properties
        Oak Ridge National Laboratory
        Oak Ridge, TN 37831

***   Chemical Sciences and Materials Systems Laboratory
****   Electrochemical Energy Research Laboratory
        General Motors R&D Center
        Warren, MI 48090

ABSTRACT

The failure stress distributions of developmental $Yb_{0.27}Co_4Sb_{12.08}$ (n-type) and $Ce_{0.86}Co_{1.02}Fe_{2.98}Sb_{11.97}$ (p-type) skutterudites were measured as a function of temperature. These thermoelectric materials are attractive candidates for use in devices under consideration for power generation sourced from intermediate to high-temperature waste heat recovery. A self-aligning, high-temperature-capable, three-point-bend fixture was developed and used to test specimens whose cross-sectional dimensions were equivalent to those typically used in thermoelectric device legs. The strength of the n-type skutterudite was approximately 115 MPa and independent of temperature to 500°C. The strength of the p-type skutterudite was equivalent to that of the n-type material and independent of temperature to at least 200°C, but its strength dropped by ~20% at 400°C. Compared to other skutterudites, the herein tested compositions have equivalent or even superior strength.

INTRODUCTION

Skutterudites are attractive thermoelectric material (TMs) candidates for use in thermoelectric devices (TDs) under consideration for power generation associated with intermediate- and high-temperature waste heat recovery. These (any) thermoelectric materials must first possess sufficient thermomechanical robustness to withstand the operational thermal gradients that activates their thermoelectric effect.

Tensile strength (and its scatter) is one indicator of that robustness. To support that assertion, Kingery's [1] thermal resistance parameter, R_{Therm}, is useful to first consider in context with TMs,

---

[1] Notice: This manuscript has been authored by UT-Battelle, LLC, under Contract No. DE-AC05-00OR22725 with the U.S. Department of Energy. The United States Government retains and the publisher, by accepting the article for publication, acknowledges that the United States Government retains a non-exclusive, paid-up, irrevocable, world-wide license to publish or reproduce the published form of this manuscript, or allow others to do so, for United States Government purposes.

$$R_{Therm} = \frac{S_{Tens}(1-\nu)\kappa}{\alpha E} \quad , \tag{1}$$

where S_{Tens} is tensile stress or tensile strength, ν is Poisson's ratio, κ is thermal conductivity, α is the coefficient of thermal expansion or CTE, and E is elastic modulus. One desires R_{Therm} to be as large as possible for improved thermomechanical resistance against effects caused by thermal gradients or thermal transients. The parameters ν, κ, α, and E are materials properties and are essentially unchangeable for any given TM under consideration. For TMs, the minimization of κ is purposely and primarily sought because that achievement improves thermoelectric efficiency. Additionally, TMs typically have a large CTE (> 10 ppm/°C). The ν for most TMs usually ranges between 0.25-0.30. The E's for TMs typically range between 50-140 GPa; however, the E is virtually unchangeable within a given class of TMs (e.g., skutterudites, TAGS, tellurides, etc.). Therefore, the intent to make the R_{Therm} for TMs as large as possible is primarily hindered by the inherently low κ and typically high CTE.

The remaining parameter in Eq. 1 is S_{Tens}. Its consideration is complicated by the fact that S_{Tens} for brittle materials, including TMs, is not necessarily a size- nor temperature-independent value. The S_{Tens} for brittle materials is a function of many parameters that are either intrinsic or extrinsic to the material. Why consider S_{Tens} and not the material's compressive strength? The S_{Tens} is anticipated to be at least one order of magnitude lower than compressive strength in TMs for the same amount of stressed material (as is the case for polycrystalline ceramics) because of the brittleness (i.e., low fracture toughness). Therefore, for conservative design, testing should focus on the measurement of a tensile strength in TMs. Of all the parameters used in the right-hand side of Eq. 1, the manufacturers of TMs can only tangibly increase R_{Therm} by increasing S_{Tens}. Thus, the valid measurement of strength in TMs, the identification of those flaws that limit strength, the active reduction of those flaw sizes to increase S_{Tens}, and appropriate strength-size-scaling of TMs in thermoelectric device design all need to be executed to ultimately achieve the highest probability of survival of a TD in service.

The primary goal of this study was to measure the S_{Tens} (via three-point-bend testing) as a function of temperature of developmental n- and p-type skutterudite materials under consideration for use in devices for power generation via waste heat recovery. Another goal was to measure strength using test coupons whose geometry closely mimics that of thermoelectric legs. Owing to the relatively small size of the test coupons, a third goal was to develop and use a test fixture that: (a) validly tests for failure stress, (b) has high temperature capability, (c) has a means of efficiently and easily promoting alignment, and (d) has a simplicity that enables the testing of large numbers of test coupons in a relatively short period of time.

EXPERIMENTAL PROCEDURE

Material Description

N-type thermoelectric materials ($Yb_{0.27}Co_4Sb_{12.08}$) were prepared on a 60 g scale by combining Co and Sb in approximately a 1:3 ratio and melting them by induction in a boron nitride crucible under an argon atmosphere. The resulting melt was combined with elemental Yb and Sb to a nominal composition of $Yb_{0.40}Co_4Sb_{12}$ in a carbon coated fused silica tube and then flame-sealed under a reduced atmosphere. The excess of Yb is a required due to the formation of Yb_2O_3 as a side product. The charge was melted at 1150°C for 5 minutes with subsequent annealing of the melt at 800°C for 1 week. The annealed melt was then milled, cold pressed, and annealed for an additional week at 800°C.

P-type materials were prepared by first alloying Co and Fe in a 1:3 ratio by arc-melting, then combining the alloy with elemental Ce and Sb with a nominal composition of $Ce_{1.05}CoFe_3Sb_{12.10}$ (the excess of Ce and Sb were necessary due to losses by vaporization) with a final mass of 60 g in a boron nitride crucible. The charge was melted by induction under Ar at 1100°C for 1 week.

The synthesis of both n- and p-type materials results in low density sintered powders, which were further milled in preparation for consolidation by spark plasma sintering (SPS). Consolidation of $Yb_{0.27}Co_4Sb_{12.08}$ proceeded by loading 12 g of milled powder into a 12.7-mm boron nitride coated die. The powder was placed under 50 MPa of uniaxial pressure under dynamic vacuum, then heated at a rate of 75°C/min to 720°C, and then soaked for an additional 2 minutes. The three consolidated pellets from this process were all greater than 98% theoretical density. The p-type material was processed in a similar manner to the n-type samples, but was heated at a rate of 75°C/min to 675°C with subsequent soaking for 2 minutes.

Prismatic bars of developmental n-type ($Yb_{0.27}Co_4Sb_{12.08}$) and p-type ($Ce_{0.86}Co_{1.02}Fe_{2.98}Sb_{11.97}$) skutterudite were prepared having nominal dimensions of 1.5 x 1.5 x 12 mm. Samples were cut from the cylindrical SPS consolidated ingots by wafering along the cylinder axis with a low speed diamond saw. Their edges were not chamfered. The electrical resistivity (ρ), Seebeck coefficient (S), and power factor for both materials are shown as a function of temperature in Fig. 1. The transport properties were evaluated using an Ulvac ZEM-3 system. The ρ and S responses for the n-type material are equivalent to responses measured previously on a similar material [2].

For the identification of strength-testing temperatures in a material whose general response is an unknown, it is often useful to first perform dilatometry to identify any temperature where a potential change of state could be occurring [3] or where the material goes into a non-equilibrium state. Dilatometry was performed on both the n- and p-type skutterudite, and the responses are shown in Fig. 2. There was no observable inflection in the elongation as a function of temperature response for the n-type material, but there was at approximately 425°C for the p-type material. This was observed on two specimens, so its response was consistent. Therefore, strength testing with the n-type was performed through 500°C and that for the p-type to 400°C.

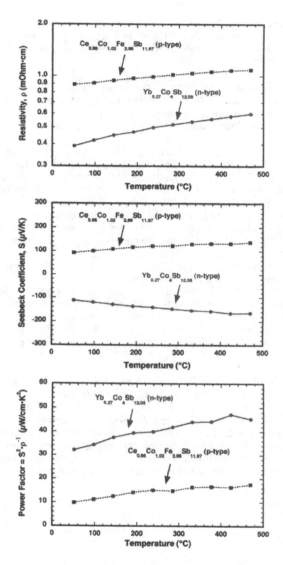

Figure 1. Electrical resistivity (top), Seebeck coefficient (middle), and power factor (bottom) as a function of temperature for the n- and p-type skutterudites.

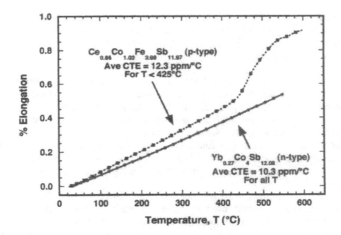

Figure 2. Percent elongation as a function of temperature for the n- and p-type skutterudites. The p-type material underwent a change of state ~ 425°C resulting in an increased rate of elongation.

Strength Testing Description: Three-Point-Bending via "Ball-on-Two-Roller Loading"

The cross-sectional dimensions of thermoelectric legs in TDs are often 1 x 1 to 4 x 4 mm, so three-point flexure strength testing was performed on prismatic bars having cross-sectional dimension in that range (1.5 x 1.5 mm in this case). This approach was deemed superior to other types of strength tests and geometries (e.g., ring-on-ring biaxial flexure testing of disk specimens) because produced failures can be caused by flaw-types and flaw-locations (e.g., edges, etc.) that are likely to exist in thermoelectric legs.

Owing to the relatively small size of the test coupons, a test fixture was conceived and developed that has high temperature capability, has a means of efficiently and easily promoting alignment, and has a simplicity that enables the testing of large numbers of test coupons in a relatively short period of time. The fixture is shown in Fig. 3. It consists of an alumina tube whose outside diameter closely fits the inside diameter of a larger tube. The inner tube has two grooves cut in one end that retain smaller alumina rollers which serve as the support span for the specimen. The specimen is set and centered on the rollers, and then an alumina sphere is set on the specimen, which provides the top or third-point loading onto the specimen. The all-alumina fixture construction enables high temperature testing. The diameter of the sphere and the outer diameter of the inner tube are matched, and are kept concentrically aligned by the inner diameter of the outer tube throughout the testing duration. An exploded view of the fixture and specimen is shown on the left of Fig. 3 and its assembly is depicted on the right. The top of the sphere is higher than the outermost tube, and it is compressively loaded by a push rod thusly loading the top of the specimen. With the specimen inside the fixture, the entire assembly is then placed inside the hot zone of a radiantly heated furnace, which is all within the working zone of an electromechanical test frame. Two complete fixtures were fabricated and the use of both enabled a high throughput rate of testing.

Another attribute of this fixture is its inherent closed assembly restricts the ambient environment from coming into contact with the specimen surface and enables testing to be done in ambient air even if the material is susceptible to oxidation. If oxidation is still operative then, alternatively, there is a "reservoir" formed by the inner tube's inner diameter that can be filled with metal filings or graphite powder which can serve as an interior oxygen getter. Using metal filings or graphite powder with the fixture was found not to be necessary with the testing of these two skutterudite compositions because the "closed assembly" by itself sufficiently suppressed oxidation of the test specimen surfaces (even to 500°C for the n-type material).

At first glance, the ratio of the test span to specimen height (8.0/1.5 or 5.33) used in the present study is much smaller than that used in other flexure tests and fixtures. For example, ASTM C1161B [4] has a span to height ratio of 40/3 or 13.33 and (the discontinued) ASTM F417 [5] has a ratio of 25.4/1.8 or 14.11. A ratio that is too small in bend testing can cause high contact stresses to occur between the rollers and specimen if a material's outer-fiber tensile stress is relatively high, and those high contact stresses can cause premature (and potentially misleading) failure. A contributing reason why the present study's ratio of 8.0/1.5 works is the three-point (or outer-fiber) failure force of these test coupons is relatively low because these materials are relatively weak, and is lower than the force necessary to initiate Hertzian cracking at either the two bottom rollers or in the vicinity of the Hertzian contact circle between the upper side of the bend specimen and the loading sphere. All flexure specimens had failure initiation occur at their tensile surface, which supports this.

Figure 3. Self-aligning, high-temperature (all-alumina) three-point-bend fixture used to measure flexure strength of 1.5 x 1.5 x 12 mm prismatic bars. Bend span is 8 mm. Exploded (left) and assembled views (right).

Producing failure initiation at the outer fiber surface is itself not a validation of testing because one must be able to accurately estimate the associated failure stress too. The maximum outer-fiber tensile stress of the developed test configuration, as estimated by finite element analysis (FEA), was compared with that calculated from the classical analytic beam bending equation. This comparison is shown in Fig. 4. Two FEA models were constructed: one representing the traditional "roller-on-two-roller" configuration and the other being the "sphere-on-two-roller" configuration used in this study. An elastic modulus of 135 GP and a Poisson's ratio of 0.20 were used for the thermoelectric materials [2]. The outer-fiber tensile stresses that they produced were identical and were within 3% of that of the classical three-point-bend equation. A 3% difference is small given the potential inaccuracies of any FEA model. Therefore, the analytic classical beam bending equation was used to estimate failure stress for all specimens.

The failure stress (S_{Tens}) was therefore calculated using the analytic three-point bending equation

$$S_{Tens} = \frac{3PS_L}{2bh^2} \ , \tag{2}$$

where P is the failure force, S_L is the support span (8.0 mm), b is the specimen base (1.5 mm nominal), and h is the specimen height (1.5 mm nominal). All testing was done at a crosshead displacement rate of 0.1 mm/min in an electromechanical test frame until fracture was produced and its associated failure load (P) recorded. For high temperature testing, the specimen/fixture assembly was soaked at temperature for approximately 10 minutes prior to the commencement of loading. Testing was performed in ambient air. As indicated earlier, there was no observable evidence of oxidation on the surfaces of the test specimens.

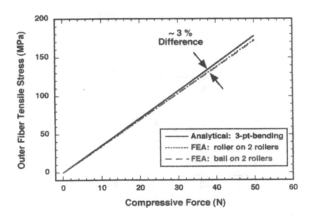

Figure 4. Comparison of the outer fiber tensile stress as a function of 3-point-bend force for the analytical case (black), its finite element analysis (blue), and the case for the bend fixture shown in Fig. 3 (red). Difference was small (~3%), so the analytical expression was used to estimate failure stresses in this study.

Commercial statistical software was used to fit the strengths to an uncensored, unimodal two-parameter Weibull distribution using maximum likelihood estimation. 95% confidence estimates were determined and used to assess any evidence of temperature-dependent changes in strength.

RESULTS AND DISCUSSION

The strength of the n-type $Yb_{0.27}Co_4Sb_{12.08}$ and p-type $Ce_{0.86}Co_{1.02}Fe_{2.98}Sb_{11.97}$ skutterudites as a function of temperature is shown in Fig. 5 and also summarized in Table I. The strength of the n-type skutterudite was independent of temperature to 500°C. The strength of the p-type skutterudite was equivalent to that of the n-type material and also was independent of temperature to at least 200°C. The p-type's strength dropped by ~20% at 400°C. As shown in Fig. 2, both skutterudite materials were believed to be in an equilibrium state at all test temperatures at which they were tested.

The values of the measured failure stresses are made in context with the Weibull effective size according to Table II, and their relationship can then be incorporated and adapted in (probabilistic) TD design. The Weibull strength data was not censored for this study, but fractography revealed that specimens failed from each of edge-, surface-, and volume-type strength-limiting flaws. Given the effective sizes listed in Table II, the characteristic strength of the n-type skutterudite was approximately 115 MPa for all temperatures, and the p-type material had that strength at 200°C and below. However, that failure stress would scale with effective size (length, area, and volume) of the thermoelectric leg according to [6]

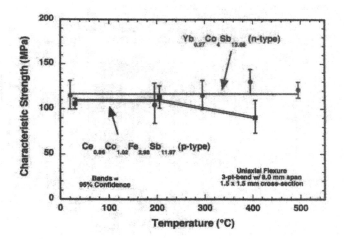

Figure 5. Characteristic strength as a function of temperature for the n- and p-type skutterudites.

Table I. Uncensored and unimodal failure stress statistics of the n- and p-type skutterudite thermoelectric materials. Three point bend testing with 8 mm span. 1.5 x 1.5 mm nominal cross-section of test specimens.

Material	Temp (°C)	# of Tests	Weibull Characteristic Strength* (MPa)	Weibull Modulus*	Gaussian Ave. Strength (MPa)	Gaussian Std. Dev. (MPa)
Yb$_{0.27}$Co$_4$Sb$_{12.08}$ (n-type)	25	17	115 (100,131)	3.6 (2.5, 5.7)	103	32
	200	15	105 (84, 128)	2.4 (1.7, 3.9)	93	39
	300	15	115 (100, 131)	3.8 (2.6, 6.1)	104	30
	400	15	130 (117, 144)	4.9 (3.4, 7.9)	120	30
	500	15	121 (112, 130)	7.3 (5.0, 11.9)	114	21
Ce$_{0.86}$Co$_{1.02}$Fe$_{2.98}$Sb$_{11.97}$ (p-type)	25	14	106 (100, 112)	9.7 (6.8, 16.0)	101	11
	200	15	113 (101, 126)	4.8 (3.4, 7.7)	104	24
	400	8	90 (73, 110)	3.4 (2.2, 7.1)	82	24

* Values in parenthesis = ± 95% confidence interval

$$P_f = 1 - \exp\left[-k_L L\left(\frac{S_{Tens}}{\sigma_{0L}}\right)^{m_L}\right] \text{ , or} \tag{3}$$

$$P_f = 1 - \exp\left[-k_A A\left(\frac{S_{Tens}}{\sigma_{0A}}\right)^{m_A}\right] \text{ , or} \tag{4}$$

$$P_f = 1 - \exp\left[-k_V V\left(\frac{S_{Tens}}{\sigma_{0V}}\right)^{m_V}\right] \text{ ,} \tag{5}$$

where P_f are probabilities of failure, $k_L L$, $k_A A$, and $k_V V$ are effective lengths, areas, and volumes, respectively and as shown in Table II, S_{Tens} is applied tensile stress, σ_{0L}, σ_{0A}, and σ_{0V} are length, area, and volume scaling parameters, respectively, and m_L, m_A, and m_V are Weibull moduli for length, area, and volume distributions, respectively. Such analysis is done numerically with life prediction software, and an example of such analysis can be found in Ref. [8].

For equivalent effective sizes, the characteristic strength of these skutterudites was stronger than both n- and p-type bismuth telluride thermoelectric materials recently tested by the authors [9].

Table II. Effective sizes for three-point bend testing. A fixture span (S_L) of 8.0 mm, square cross-section ($b = h$) 1.5 x 1.5 mm, and a Weibull modulus (m) = 5 used in the calculations. Failure stress in these materials can be limited by edge-, surface-, or volume-type strength-limiting flaws.

Effective Size	Equation	for $m_i = 5$
Effective Length $L_{eff} = k_L \cdot L$	$(2 \cdot S_L) / (m_L + 1)$	2.67 mm
Effective Area $A_{eff} = k_A \cdot A$	$2 \cdot S_L \cdot (h+b) \cdot (m_A+2) / (4 \cdot (m_A+1)^2)$ [7]	2.33 mm^2
Effective Volume $V_{eff} = k_V \cdot V$	$b \cdot h \cdot S_L / 2 \cdot (m_V+1)^2$ [7]	0.25 mm^3

But more relevantly, compared to other strength-related literature on skutterudites, the strength of the herein described skutterudites are equivalent or superior. Electrical discharge machined prismatic bars of 1.5 x 2.0 x 28 mm of n-type $CoSb_3$ and p-type $CeFe_{3-x}Ru_xSb_3$ were flexure strength tested at room temperature in Ref. [10]. The authors cite the general use of ASTM C1161 [4] and 4-point-bending but do not specifically state the flexure fixture size they used though it is surmised here they used 10 and 20 mm spans. They report a characteristic strength of 91 MPa for the n-type material, and 40 MPa for the p-type material. To enable a comparison to the present study's strength statistics in Table II (i.e., m = 5), their results produce associated $k_L L = 23.3$ mm, $k_A A = 26.3$ mm^2, and $k_V V = 2.9$ mm^3 (see Ref. [7] for $k_A A$ and $k_V V$ calculations for 4-point-bending). Using the general Weibull scaling form of S_2 = (effective size 1 / effective size 2)$^{1/m} \cdot S_1$ for the same probability of survival shows the present study's failure stresses would expected to be ~ 50-60% higher. This suggests that the strength of the n-type materials in the two studies is similar but the p-type material strength in the present study is superior.

SUMMARY AND CONCLUSIONS

The failure stress distributions of $Yb_{0.27}Co_4Sb_{12.08}$ (n-type) and $Ce_{0.86}Co_{1.02}Fe_{2.98}Sb_{11.97}$ (p-type) skutterudites were measured as a function of temperature. A self-aligning, high-temperature-capable three-point-bend fixture was developed and used to test specimens whose cross-sectional dimensions (1.5x1.5 mm) are equivalent to those typically used in thermoelectric device legs. Tests were run at sufficiently rapid loading rates so fast-fracture or inert strengths were measured.

The strength of the n-type skutterudite was approximately 115 MPa ($L_{eff} = 2.67$ mm or $A_{eff} = 2.33$ mm^2 or Veff = 0.25 mm^3 for m = 5), and that was independent of temperature to 500°C. The strength of the p-type skutterudite was equivalent to that of the n-type material (same effective sizes) and also was independent of temperature to at least 200°C. The p-type's strength dropped by ~20% at 400°C. The herein described n-type skutterudite has a strength comparable to another n-type skutterudite in the literature but this study's p-type has a superior strength.

ACKNOWLEDGEMENTS

Research sponsored by the U.S. Department of Energy, Assistant Secretary for Energy Efficiency and Renewable Energy, Office of Vehicle Technologies, as part of the Propulsion Materials Program, under contract DE-AC05-00OR22725 with UT-Battelle, LLC. The authors thank the University of Wisconsin-Platteville's O. M. Jadaan for assistance with the effective size analysis, and ORNL's F. Ren and J. -A. Wang for reviewing the manuscript. JRS and JY would like to thank J. F. Herbst and M. W. Verbrugge for their continued support and encouragement.

REFERENCES

[1]   W. D. Kingery, "Factors Affecting Thermal Stress Resistance of Ceramic Materials," *Journal of the American Ceramic Society*, 38:3-15 (1955).

[2]   J. R. Salvador, J. Yang, X. Shi, H. Wang, A. A. Wereszczak, H. Kong, and C. Uher, "Transport and Mechanical Properties of Yb-Filled Skutterudites," *Philosophical Magazine*, No. 19, 89:1517-1534 (2009).

[3]   A. A. Wereszczak, H. -T. Lin, T. P. Kirkland, M. J. Andrews, and S. K. Lee, "Strength and Dynamic Fatigue of Silicon Nitride at Intermediate Temperatures," *Journal of Materials Science,* 37 1-16 (2002).

[4]   "Standard Test Method for Flexural Strength of Advanced Ceramics at Ambient Temperature," ASTM C 1161, Vol. 15.01, ASTM, West Conshohocken, PA, 2008.

[5]   "Test Method for Flexural Strength (Modulus of Rupture) of Electronic-Grade Ceramics (Withdrawn 2001)," ASTM F417, ASTM, West Conshohocken, PA, 1996.

[6]   "Standard Practice for Reporting Uniaxial Strength Data and Estimating Weibull Distribution Parameters for Advanced Ceramics," ASTM C 1239, Vol. 15.01, ASTM, West Conshohocken, PA, 2008.

[7]   G. D. Quinn, "Weibull Strength Scaling for Standardized Rectangular Flexure Specimens," Journal of the American Ceramic Society, 86:508-510 (2003).

[8]   O. M. Jadaan and A. A. Wereszczak, "Probabilistic Design Optimization and Reliability Assessment of High Temperature Thermoelectric Devices," *Ceramic Engineering and Science Proceedings*, [3], 29:157-172 (2008).

[9]   A. A. Wereszczak, T. P. Kirkland, O. M. Jadaan, and H. Wang, "Strength of Bismuth Telluride," *Ceramic Engineering and Science Proceedings*, 30, [9] 131-142 (2009).

[10] V. Ravi, S. Firdosy, T. Caillat, B. Lerch, A. Calamino, R. Pawlik, M. Nathal, A. Sechrist, J. Buchhalter, and S. Nutt, "Mechanical Properties of Thermoelectric Skutterudites," pp. 656-662 in CP969, Space Technology and Applications International Forum – STAIF 2008, Ed. M. S. El-Genk, American Institute of Physics, 2008.

GRAPHITE AND CERAMIC COATED PARTICLES FOR THE HTR

Heinz Nabielek[i], Mark Mitchell[ii]
i) Inst. Energy Research, Forschungszentrum Jülich GmbH, Jülich, Germany
ii) Pebble Bed Modular Reactor (Pty) Ltd, Centurion, South Africa

ABSTRACT

The fully ceramic core of a High-Temperature Gas-Cooled Reactor (HTR) consists of graphite internals and graphitic fuel elements that contain coated particles. These are fissile kernels that are surrounded by several coating layers: in the TRISO (tri-isotropic coated) particle there is a SiC interlayer interspersed between two dense pyrocarbon layers. While coatings originally had only been introduced for the handling of the fuel, they have been optimized in the meantime to withstand operational gas pressures, to retain fission products generated during irradiation and to keep them inside in accidents. The latest development showed the dense coating layers also to be efficient in long-term storage of HTR fuel.

Graphite is used for the core structure of HTRs as a stable and strong material, withstanding temperatures of well over 2000°C and a complete lack of interaction with the fuel and with the helium coolant.

INTRODUCTION

In HTRs, safety is ensured not only by active safety systems that must act to protect the core, but by a large negative temperature coefficient and the passive removal of heat generated by the core even with a complete loss of coolant. This inherent safety leads to a reduction in the requirement for active safety systems that must be guaranteed to operate when required to protect the core. The intention is that this reduction will provide the HTR design with a significant economic and safety advantage and enable them to compete economically with modern low-cost power generation.

The design of modern small modular HTR aims at a reactor core that is inherently safe. The concept of inherent safety is defined by the following features[1]:

> the nuclear power of the reactor and the fuel element temperature are limited in a self-acting manner,
> the temperatures of the fuel elements are limited even in the most severe accident conditions by means of passive decay heat removal,
> chemically stable materials limit destruction by chemical attack, and
> the reactor core maintains its geometry and the cooling and shut-down functions under all foreseeable accident conditions.

The construction of the core internals from graphite is significant for the design of the reactor, specifically ensuring that even during a worst case accident:

> the reflectors form part of the passive heat removal path which ensures that fuel temperatures do not exceed their limits, and

the reflectors maintain sufficient structural integrity to ensure that the geometry of the core is unchanged and that the absorber material used for control and shutdown of the nuclear chain reaction can be introduced.

The HTR is the logical consequence of improvements of gas-cooled graphite moderated reactors based on the earlier British development of Magnox[2] and AGR[3] (Advanced Gas-cooled Reactors with CO_2 coolant).

GRAPHITE

The graphite material in the core is subject to severe operating conditions: typically, in the lifetime of a pebble-bed modular reactor[4], the reflector graphite is exposed to fast neutron fluence levels in excess of $1.5x10^{26}$ neutron/m^2 in Dido-Nickel dose EDN (equivalent to 20 dpa) at temperatures of approximately 700 C. In accidents, small parts of the graphite reflector may reach temperatures near 1600°C. These conditions induce changes in the material that must be considered in the design of the reactor.

The most significant targets for ideal reflector graphite for nuclear applications are summarized by EPRI[5]. These values are based on previous experience gained in the manufacture and operation of nuclear grade graphite for use in core internals of Magnox, AGR and HTR reactors. These material property targets are summarized below in Table 1.

Table 1: Requirements for an Ideal Reflector Graphite[5,6]		
Property	Required Range	Reason
Density	~1.8 g/cm^3	Density shows that the material is not excessively porous and also provides for an effective neutron reflection/ shielding per unit volume. High density also indicates high strength.
Thermal conductivity at room temperature	~145 W/m/K	Indicates a good degree of graphitization and guarantees good heat transport.
Low absorption cross-section	~4 mbarn	Required for neutron efficiency of the core. The limiting neutron absorption is that of pure carbon (~3.5 mbarn).
Impurity levels (total ash content)	< 250 ppm	Required to minimize activation and reduce susceptibility to oxidation.
Coefficient of thermal expansion (20°C to 120°C)	$4.0-5.5x10^{-6}$ K^{-1}	This low value indicates that the graphite will have a high dimensional stability when subjected to fast neutron irradiation and thermal cycles.
Strength	~20 MPa	Reasonable strength is required to allow for the use of the graphite as a structural component.
Anisotropy	< 1.15	Measure of how isotropic the graphite is. The more isotropic, the more suited is the material for use in a nuclear reactor.

GRAPHITE MATERIAL DEVELOPMENT AND CHARACTERIZATION

The development and characterization of suitable nuclear grade graphite has the following objectives:

Develop a graphite grade and establish a production route for the manufacture.

Qualify the production route, the necessary quality assurance and quality control measures.

Characterize the material to be used to provide adequate design data - both for unirradiated and irradiated materials.

MATERIAL SELECTION AND QUALIFICATION

The material grade that was developed for use in PBMR, the South African pebble-bed reactor, is NBG-18 as manufactured by SGL Carbon GmbH in Germany. This material is a medium grained, vibration molded, isotropic material as described in the ASTM specification for nuclear graphite[7]. Extensive trial production has been completed to qualify the production process and ensure that the material can be produced consistently in future.

MATERIAL CHARACTERIZATION – DESIGN INPUT DATA

Graphite input data needed in the design are summarized in Table 2. These properties are typical as defined by the applicable standards committees[8,9].

Table 2: Properties of Graphite to be determined as Design Information Data						
Property	Irradiation dependence	Field of application				
		Neutron Physics	Thermo-dynamics	Design	Stress Analysis	Radiology
Density		X	X	X		
Capture Cross Section		X				
Thermal Conductivity	X		X			
Specific Heat Capacity			X			
Thermal Expansion	X			X		
Strength	X			X		
Oxidation Behavior				X	X	
Young's Modulus	X				X	
Poisson Ratio of Elasticity					X	
Irradiation induced dimensional change	X				X	
Irradiation induced creep	X				X	
Detailed Chemical Analysis						X

Typical unirradiated properties, compared to the properties of historical graphite grades, are shown in Table 3.

The determination of the effect of fast neutron irradiation on the material presents a challenge. To allow for the design to proceed a model that predicts the irradiation induced behavior of the material was generated, based on historical ATR-2E and VQMB data. This model will be verified by completing an irradiation test program. Knowledge of the irradiation induced dimensional change strains (also referred to a Wigner strains), irradiation induced creep and changes to the other properties are key to determining the operating life of the reactor. The model represents the properties identified as irradiation dependent in Table 2. These properties will be measured as part of the irradiation test program. An example of the effect of fast neutron irradiation on a graphite grade[10] is shown in Figure 1: it can be seen that shrinkage is in the order of 3% and useful lifetime is in around 35 to 50 dpa at temperatures around 300°C.

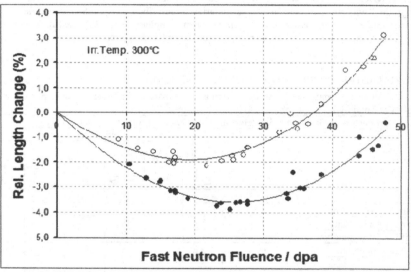

Fig. 1: Dimensional changes due to fast neutron irradiation at 300°C (white data points: perpendicular; black data points: parallel to preferential orientation) of graphite grade ATR-2E[12].

Table 3: Comparison of measured NBG-18 properties to the properties of other nuclear graphite grades

GRADE		ATR-2E	VQMB	H-451	IG-110[a]	PGX[a]	P2XA2N[a]	NBG-18[b]
Manufacturer		SIGRI[c]	UCAR[d]	SGL[e]	Toyo-Tanso[f]	UCAR	SGL	SGL
Filler Coke		Rutgers Pitch	Pitch	Petro-Coke	Petro-Coke	Petro-Coke	Pitch-Coke	Pitch-Coke
Forming Method		Extruded	Molded	Extruded	Isopressed	Molded	Extruded	Vibe Molded
Impregnation		2x	2x	2x	2x	2x	2x	2x
Bulk Density (g/cm³)		1.77 – 1.80	1.86	1.74	1.77	1.73	1.81	1.87
CTE (10^{-6}.K^{-1})	L	4.4 – 4.7	4.1	3.3	4.0	2.6	5.1	4.49
	T	4.9 - 5.5	4.3	4.0	3.6	2.2	5.2	4.62
Thermal Conductivity (W.m^{-1}.K^{-1})	L	121- 179	120	163	124	104	135	145
	T	107 - 163	-	149	138	125	-	140
Young's Modulus (GPa)	L	9.6 – 11.4	11.0	10.6	9.42	5.5	9.6	11.9
	T	8.4 – 9.7	10.1	9.55	9.97	7.6	10.4	11.6
Tensile Strength (MPa)	L	12.6 – 15.5	16.9	14.2	24.9	9.3	17	21
	T	12.4 – 12.5	13.1	13.9	24.0	10.0	20	20
Compressive Strength (MPa)	L	52.0 - 55.9	58.9	--	78.4	30	65	77.4
	T	52.9 - 57.8	56.2			31	76	78.5
Flexural Strength (MPa)	L	23.0 – 26.6	22.4	--	39.2	10.5	27	31
	T	18.9 – 21.0	19.9			12	33	30
Ash Content (ppm)		170 - 1120	< 1000	50	20	70	120	< 250
Neutron Cross-Section (mbarn)		< 5.0	4.3	-	-	-	-	< 5.0
Isotropy Ratio CTE$_{max}$/CTE$_{min}$		1.11 – 1.16	1.10	1.21	1.11	1.18	1.03	1.03

Note: L/T = longitudinal/transverse direction, parallel/perpendicular to extrusion/pressing direction, or, an orthogonal direction

[a]Manufacturer's values; [b]Actual pre-production data; [c]German; [d]USA; [e]French-German; [f]Japanese; [g]Estimated value;

COATED PARTICLE FUEL

The coated particle consists of a UO_2 fuel kernel, a porous buffer layer and a series of dense isotropic layers: inner pyrocarbon, SiC and outer pyrocarbon; designated a TRISO (=tri-isotropic) particle. These layers are mechanically strong and retain fission products throughout irradiation within the limits for HTR typical operation temperatures und burnup.

There is a clear set of standards for modern high quality fuel[11] in terms of low levels of heavy metal contamination, manufacture-induced particle defects during fuel body and fuel element making, irradiation/accident induced particle failures and limits on fission product release from intact particles.

While HTR design is still open-ended with blocks for the prismatic and spherical fuel elements for the pebble-bed design, there is near world-wide agreement on high quality fuel: a 500 μm diameter UO_2 kernel of 10% enrichment is surrounded by a thick sacrificial buffer layer to be followed by a dense inner pyrocarbon layer, a high quality silicon carbide layer of theoretical density and another dense outer pyrocarbon layer. The following values for thickness and density represent typical coating layer properties:

- buffer pyrocarbon layer   95 μm   1.0 g/cm^3
- inner pyrolytic carbon layer   40 μm   1.9 g/cm^3
- silicon carbide layer   35 μm   3.20 g/cm^3
- outer pyrolytic carbon layer   40 μm   1.9 g/cm^3 .

Good performance has been demonstrated both under operational conditions to 12% FIMA (= Fissions per Initial heavy Metal Atoms) and more and under accident conditions to a maximum 1600°C.

A variety of coating failure mechanisms[12] has been analyzed and studied in past decades to understand and quantify the mechanical and chemical behavior of the compound structure of the coated particles under operational and accidental conditions. The most elementary in-reactor effect is the build-up of gas pressure in the free volume of the porous buffer layer inducing circumferential stresses that ultimately may exceed the tensile strength of the SiC leading to the release of fission products. Major emphasis is therefore placed in the determination of the release of stable and long-lived Xe and Kr as the indicator of a coating failure and the formation of CO. Both strength and strength distribution of SiC and the pyrocarbon layers are the most important material input parameters to enable failure predictions with fuel performance codes [13,14,15,16].

Recent interest in high burn-up fuel for better economy and waste reduction and coated particles that can operate at very high temperatures (e.g., hydrogen production for fuel cells) will need continued modeling work and code development. Scoping calculations with the code PA-NAMA[15,17] have been useful for exploring the operating range of the fuel up to 20% FIMA.

The potential failure mechanisms including statistics on manufacturing variations, extreme irradiation conditions leading to kernel migration and fission product attack due to accident conditions like core heatup, water ingress, air ingress and others have to be regarded. The coating layers retain their strength and fission product retentiveness in accidents limited to 1600°C. The HTR core is designed such that temperatures are restricted to keep particles intact and retentive[18,19,20].

HTR FUEL DESCRIPTION

TRISO coated fuel particles imbedded in graphite form the basis of fuel elements utilized in most high temperature reactor designs. Modern HTR fuel is designed around the following envelope[21]:

maximum core temperatures of 1200°C in operation and 1600°C in accidents;

maximum burnup of 12% FIMA, potential for 20% FIMA;

maximum fast neutron fluence of 6×10^{25} m^{-2} (E>16 fJ);

maximum power densities of 12 MW/m^3;

Fuel design criteria are derived from the fundamental safety functions and from the functions that the fuel must perform in the environment in which it will be used. The main criteria can be described as follows:

Fuel integrity must always be maintained throughout the whole life cycle of the fuel.

Fuel must be able to withstand transport and handling stresses during shipping and handling, and remain intact under all expected and design reactor conditions.

Fuel must contain fission products for the entire life cycle of the fuel.

To achieve these criteria, coated particles failure fractions must remain low enough not to cause any significant radiological risk to operating personnel and the general public. During irradiation, fuel integrity is observed by monitoring gaseous fission product release from the fuel in the primary coolant.

FUEL DESIGNS

All HTR fuel designs can be described as either prismatic or spherical fuel designs. Various prismatic fuel designs have been developed, in particular, in the USA and Japan (Figure 2). Modern spherical fuel design is based on the German reference fuel for the HTR-MODUL (High Temperature Reactor – Modular)[22] and is presented in Figure 3. This fuel design was extensively tested and evaluated in Germany and the Netherlands. Also, spherical fuel elements were tested on a large scale during 21 years of AVR[23] operation at Jülich, and in the 300 MW$_e$ THTR reactor, both in Germany[24].

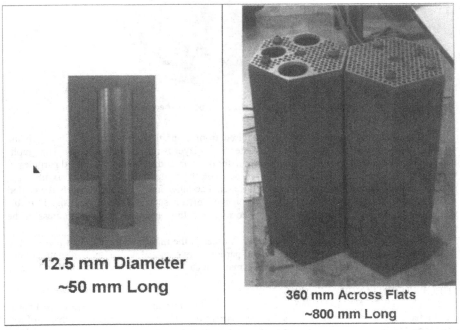

12.5 mm Diameter
~50 mm Long

360 mm Across Flats
~800 mm Long

Fig. 2 Typical US prismatic fuel element design: compact left, graphite fuel block right[25].

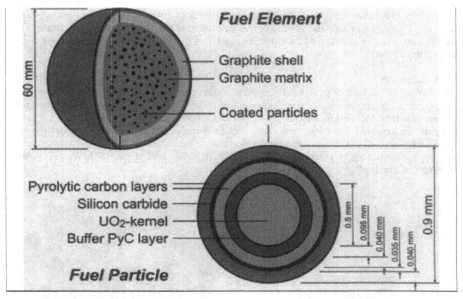

<u>Fig. 3:</u> Spherical fuel element and coated particle for pebble-bed reactors[26],[27].

Spherical fuel elements are manufactured from graphite matrix material, in which the TRISO coated particles are imbedded. The outer 5 mm layer is matrix material only. The graphite matrix material functions as a good heat transfer medium and stabilizes the coated particles in the sphere. Good thermal contact is achieved between the coated particles and matrix material, so that low temperature gradients occur in the fuel. The outer fuel free zone protects the coated particles from damage from outside direct mechanical effects such as abrasion and shock. It further acts as a barrier layer against chemical corrosion in the case of water or air ingress in the core.

To prevent coated particles to touch each other in the matrix material which may lead to failures during the pressing stage, each coated particle is overcoated with a layer of matrix material graphite before being mixed with the bulk matrix material.

FUEL PERFORMANCE

The code PANAMA[15] simulates the mechanical performance of TRISO coated fuel particles under given normal operation and accident conditions. The failure probability, that is of importance under the conditions of normal reactor operation and core heatup accidents for modular type HTRs, is based on a pressure vessel model and includes degradation effects on the SiC layer due to fission product corrosion.

The pressure vessel is assumed to fail as soon as the stress induced in the SiC layer by the internal gas pressure exceeds the ultimate tensile strength of silicon carbide. The probability for a pressure vessel failure of a particle is a function of time and temperature and can be described by means of a Weibull relationship.

The tensile strength is a material parameter whose median value and modulus is derived from dedicated experiments like the Bongartz brittle ring test[28,29,30,31,32]. SiC strength is weakened under irradiation; both median strength and modulus decrease with neutron fluence at irradiation above 1000°C. The effect of fission product corrosion is transformed into an effective thinning of the SiC layer, representing the pressure vessel wall, at a volume corrosion rate according to Montgomery[33], thus leading to a sooner failure of the coated particle at given conditions.

The internal gas pressure is calculated by applying the gas state law to the generation of fission gases Xe, Kr, and reaction gas CO. The amount depends on the yield of stable fission gases, the burnup, the number of oxygen atoms produced in the kernel, and the temperature and irradiation time. Oxygen production in the particle kernels as a result of the fissioning of ^{235}U or ^{239}Pu is strongly dependent on the irradiation history and to a great extent on the type of kernel. Corresponding relationships for CO/CO_2 molecules generated were derived from tests at Seibersdorf, covering an irradiation time up to 550 days and a temperature range between 950-1525°C[34].

The validation of the PANAMA model has been made against numerous experiments with spherical fuel elements irradiated and also heated at accident temperatures in the range of 1600-2500°C[35,36,37]. Good agreement between predicted particle failure and experimentally derived failure from Kr release measurements had been achieved.

FINAL DISPOSAL BEHAVIOR OF HTR FUEL

Direct disposal of spent spherical HTR fuel elements is an attractive option, because fuel and fission products are already well contained. It is the reference concept for the management of the 1 million fuel elements from the operation of the two German high temperature reactors AVR and THTR. The demonstration program investigates the long term safety under final disposal conditions and determines the radionuclide release in case of contact with disposal site relevant aqueous phases.

Several investigations show that the long term safety is related to intact coating layers. The porous graphite matrix may allow access of water that comes into contact with coated particles. For the long-term storage of spent HTR fuel we need to know, if the containment function of the particle coating layers remain effective over long periods of time.

The HTR fuel element is a complex multibarrier system with respect to long term radionuclide mobilization under final disposal conditions[38,39]. Each barrier contributes to the immobilization of the radionuclides in a final repository. To develop a model for long term performance calculations it is necessary to understand the behavior of each component. The fuel kernel of UO_2 or $(Th,U)O_2$ is a confinement matrix for a large fraction of generated fission products and actinides. Since this matrix is stable to dissolution in water, it constitutes a barrier against radionuclide release upon ground water access. The matrix does not confine a certain fraction of radionuclides such as fission gases xenon and krypton, cesium and iodine in the fuel kernel, because they segregate to grain boundaries and buffer porosity.

The fuel kernels are surrounded by dense and water-resistant coating layers that have to be disintegrated before an aqueous phase could come into contact with the fuel kernel. Therefore the coating layers can be regarded as another most effective barrier.

The graphite matrix of the fuel element will restrict water access to the coated particles and represents the outer barrier.

The radionuclides in the graphite matrix may come into contact with aqueous phases, if the groundwater penetrates the storage cask. However, the radiotoxic inventory of the graphite matrix is negligibly small in comparison to the inventory in the fuel kernels.

For the fuel kernels, two cases have to be considered: a few fuel kernels with a defective or failed coating and the major number of kernels (> 99 %) with an intact coating. The access of aqueous phases to the failed particles will be controlled by the pore system of the graphite matrix by a diffusion process. Radionuclides in the defective kernels, which were segregated to grain boundaries, will probably be rapidly dissolved in the pore fluid and migrate slowly by diffusion to the surface of the fuel pebble. The major fraction of the radionuclide inventories of these kernels is still retained and they are dissolved only very slowly as determined by the dissolution rate of the fuel kernel matrix.

Most fuel kernels are protected from water access by the coatings, which consist of different layers. The BISO particle has a porous buffer layer and a dense pyrocarbon layer. The TRISO particle has an additional SiC interlayer in between the dense pyrocarbon layers. A release of radionuclides from segregated phases and from the fuel matrix will only occur, when an aqueous medium penetrates these layers.

Each of these barriers has its own properties and behaves differently in aqueous phases. In order to obtain a long-term prediction for behavior during final disposal, it is necessary to investigate the different barrier materials and construct a model for the whole fuel pebble. The main processes are given in Figure 4.

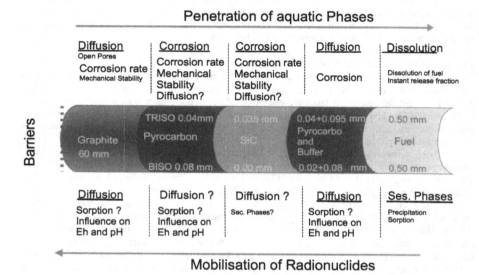

Fig. 4: Main processes for radionuclide mobilization from a HTR fuel element[38]

Leaching tests with irradiated spherical fuel elements show two clear-cut phenomena with respect to retention of ^{137}Cs and ^{154}Eu (Figure 4):

Intact particles retain the fission products nearly completely;

Release from defective/failed particles and from the sphere matrix is fast, but this contribution is small.

These results of leaching experiments are consistent with the hypothesis that mobile radionuclides in the graphite matrix will be released fast and that the main inventory is retained by the intact coated particles. All the results obtained so far show – as was expected – that the SiC layer represents the most important long term barrier of the coating materials.

CONCLUSION

Stability and fission product retention has been demonstrated for operating and accident conditions for current HTR designs. Further potential to higher operating temperatures and higher burnup are being explored.

Both in normal operation, in accidents and in long-term storage of HTR fuel, we have to quantify the protective functions of the coating layer on every single particle rather than large external barriers and containments.

Under final disposal conditions, HTR fuel is well packaged: only the radionuclide inventory of defective or failed coated particles can be mobilized. The main part of the radionuclide inventory is sealed by the coating layers. The life time of the SiC coating can be assumed to last 1000 to 10,000 years. Investigations are on-going for the long term interaction between SiC and fission products.

After successful application of nuclear graphite in Magnox, AGR, AVR, Peach Bottom, Fort St. Vrain and THTR, there is a massive amount of scientific and technological background for graphites in new HTRs. Graphite manufacturers can develop and produce successor materials with similar or better properties than legacy grades. Reactor engineers and designers can use much of the information and data available from earlier programs, but testing and analysis of the new grades is still necessary.

REFERENCES

[1] Haag, G., Kugeler, K., and Phlippen, P.W., "The HTR and the New German Safety Concept for Future Nuclear Power Plants" in IAEA TECDOC-901, Bath (United Kingdom). 24-27 Sep 1995.

[2] http://en.wikipedia.org/wiki/Magnox

[3] "Manufacturing AGR Fuel" Springfields Fuels Ltd, Springfields, Salwick Preston, Lancashire PR4 0XJ UK.

[4] Mitchell, M. N. ,"The Design of the PBMR Core Structures", Fifth International Nuclear Graphite Specialists Meeting . 12-9-2004. Plas Tan-Y-Bwlch, Maentwrog, Gwynedd, United Kingdom.

[5] EPRI, "Graphite for High-Temperature Reactors", EPRI, Palo Alto, 2001.

[6] Mitchell, M.N., "Identification of the Key Parameters Defining the Life of Graphite Core Components", SMiRT 18-C05-1, Beijing, China, 7-12 August 2005.

[7] ASTM D7219-05. Standard Specification for Isotropic and Near-isotropic Nuclear Graphites. 2008. American Society of Testing and Materials.

[8] KTA 1992, KTA-3232 Keramische Einbauten in HTR-Reaktordruckbehältern, Sicherheitstechnische Regel des KTA, KTA-3232

[9] LD-1097. Qualification Requirements for the Core Structures Ceramics of the Pebble Bed Modular Reactor. National Nuclear Regulator (NNR), Pretoria South Africa.

[10] Haag, G., "Properties of ATR-2E Graphite and Property Changes due to Fast Neutron Irradiation", Forschungszentrum Jülich report Juel-4183, Jülich, October 2005.

[11] Nabielek, H., et al. "Development of Advanced HTR Fuel Elements", Nucl. Eng. Des. 121 (1990), 199.

[12] Petti, D., "Key Differences in the Fabrication, Irradiation and High Temperature Accident Testing of US and German TRISO-Coated Particle Fuel, and their Implications on Fuel Performance", Nucl. Eng. Des. 222 (2003), 281.

[13] Verfondern, K. et al., "PANAMA-Ein Rechenprogramm zur Vorhersage des Partikelbruchanteils von TRISO Partikeln unter Störfallbedingungen", Forschungszentrum Jülich report Juel-Spez-298, February 1985.

[14] Phélip, M. et al., "The ATLAS HTR Fuel Simulation Code: Objectives, Description and First Results", 2nd Int Topical Meeting on HTR Technology HTR2004, Beijing, China, 22-24 October 2004.

[15] Miller, G.K. et al., "Current Capabilities of the Fuel Performance Modeling Code PARFUME", 2nd Int Topical Meeting on HTR Technology HTR2004, Beijing, China, 22-24 October 2004.

[16] Miller, G.K., Petti, D.A., Maki, J.T., and Knudson, D.L., "PARFUME Theory and Model Basis Report" Idaho National Laboratory Report INL/EXT-08-14497, Idaho Falls September 2009.

[17] Nabielek, H., and Verfondern, K., "Mechanical Coated Particle Failure for PuMA Fuel Design" EU Project PuMA Final Report on Pu and Minor Actinide Fuel for HTRs, Forschungszentrum Jülich, August 2009.

[18] Nabielek, H. et al. "The Performance of High-Temperature Reactor Fuel Particles at Extreme Temperatures", Nucl. Techn. 84 (1989), 62.

[19] Schenk, W. et al., "Fuel Accident Performance Testing for Small HTRs", J. Nucl. Mat. 171 (1990), 19.

[20] van der Merwe, J.J., et al. "A Method to Evaluate Fission Gas Release during Irradiation Testing of Spherical Fuel", J. Eng. Gas Turbines Power 131 (2009), 52902.

[21] van der Merwe, J.J., et al. "HTR Fuel Design, Qualification and Analyses at PBMR", PHYSOR 2006, Vancouver, Canada (2006).

[22] Reutler, H., et al., "Advantages of Going Modular in HTRs", Nucl. Eng. Des. 78 (1984), 129.

[23] VDI-Society for Energy Technologies, "AVR Experimental High Temperature Reactor: 21 Years of Successful Operation for a Future Energy Technology", VDI-Verlag, Düsseldorf 1990.

[24] Nickel, H. et al., "Long Time Experience with the Development of HTR Fuel Elements in Germany", Nucl. Eng. Des. 217 (2002), 141.

[25] Fortescue, P., F.R. Bell and R.B. Duffield, "Hexagonal Fuel Element", U.S. Patent Application No. 485,811, Filed September 8, 1965.

[26] Hrovat, M., Nickel, H., Koizlik, K., "Über die Entwicklung eines Matrixmaterials zur Herstellung gepresster Brennelemente für Hochtemperatur-Reaktoren, FZJ-Report Juel-969, (June 1973).

[27] Becker, H.-J., et al., "Method of Manufacturing Spherical Fuel Elements", European Patent EP0176727, Publication Date: 07/13/1988, Filing Date: 08/17/1985.

[28] Bongartz, K. et al. "Brittle Ring Test--a Method for Measuring Strength and Young's Modulus on Coatings of HTR Fuel Particles", J. Nucl. Mat. 62 (1976), 123.

[29] Snead, L.L. et al. "Handbook of SiC Properties for Fuel Performance Modeling", J. Nucl. Mat. 371 (2007), 329.

[30] Byun, T.S. et al., "Fracture Stress of SiC Layer in Coated Particle Fuel", Trans. Am. Nucl. Soc. 94 (2006), 680.

[31] Byun, T.S. et al., "Evaluation of Fracture Stress for the SiC Layer of TRISO-Coated Fuel Particles Using a Modified Crush Test Method", Int. J. Applied Ceramic Technology, November 2009.

[32] van Rooyen, G.T., et al. " The fracture strength of TRISO-coated particles by compression testing between soft anvils", J. Nucl. Mat., in preparation, 2010.

[33] Montgomery, F., "Fission Product SiC Reaction in HTGR Fuel", Report GA-905837, General Atomics, San Diego, USA (1981).

[34] Proksch, E. et al., "Production of Carbon Monoxide during Burnup of UO_2 Kernelled HTR Fuel Particles", J. Nucl, Mat. <u>107</u> (1982), 280.

[35] Schenk, W. et al., "Fission Product Release Profiles from Spherical HTR Fuel Elements at Accident Temperatures", Forschungszentrum Jülich report Juel-2234, Jülich, September 1988.

[36] Schenk, W. et al., "Kugelbrennelemente mit TRISO Partikeln bei Störfalltemperaturen", Forschungszentrum Jülich report Juel-Spez-487, Jülich, January 1989.

[37] Freis, D., "Störfallsimulationen und Nachbestrahlungsuntersuchungen an kugelförmigen Brennelementen für Hochtemperaturreaktoren", PhD Thesis, RWTH Aachen 2009.

[38] Fachinger, J. et al., "Behavior of Spent HTR Fuel Elements in Aquatic Phases of Repository Host Rock Formations", Nucl. Eng. Des. <u>236</u> (2006), 543.

[39] Nabielek, H., et al., "Ceramic Coated Particles for Safe Operation in HTRs and in Long-Term Storage", pp 193-202 of Ceramics in Nuclear Applications, ed. Y. Katoh et al., Wiley InterScience, 14 Jan 2010.

DEVELOPMENT AND CHARACTERIZATION OF HIGH CONDUCTIVITY GRAPHITE FOAMS FOR THERMAL MANAGEMENT APPLICATIONS

A.L. Gyekenyesi[1], M. Singh[1], C.E. Smith[1], P.G. Stansberry[2], M. K. Alam[3], and D.L. Vrable[4]

[1]Ohio Aerospace Institute, 22800 Cedar Point Road, Cleveland, Ohio 44142 USA
[2]GrafTech International Holdings, Inc., 12900 Snow Road, Parma, OH 44130 USA
[3]Department of Mechanical Engineering, Ohio University, Athens, OH 45701 USA
[4]Thermal Management & Materials Technology, 4664 Vista de la Tierra, Del Mar, CA 92014

ABSTRACT

Graphitic carbon foams are excellent candidate materials for thermal management applications due to their extraordinarily high ligament conductivity and large surface area. Graphite foams have reported ligament conductivities greater than 1800 W/m·K and bulk values up to 245 W/m·K. But before these foams can be fully utilized, numerous issues need to be studied and resolved. Improvements need to be made regarding the overall foam strength; flaking and pressure drop in forced convection systems; bonding; environmental durability; as well as cost, product size, quality and production capacity. A team comprised of the Ohio Aerospace Institute, GrafTech International Holdings, Inc., Ohio University, and TMMT, Inc. is attending to the above needs by conducting cooperative research to resolve some of the issues mentioned. The categories of research include the development of new/modified manufacturing procedures for optimizing the morphology of the graphite foam while increasing the quality and quantity; the study of various coatings for increased durability and functionality; the investigation of joining and integration technologies to allow for optimal system integration regarding strength and thermal conductivity; as well as extensive modeling addressing mechanical, thermal and fluid flow simulations/predictions from the material to system levels (multi-scale modeling). This paper offers additional project details as well as provides some preliminary results.

INTRODUCTION

Due to increased performance in a wide range of engineered products ranging from computer processors to advanced aerospace vehicles, there is a critical need for improved thermal management systems for transferring heat. The required enhancements include increased thermal conductivity, increased surface area, reduced weight/volume, as well as operability in harsh environments (e.g., durability under high flow rates, vibrations, stress, elevated temperatures, and oxidative environments). Specifically, improved thermal management is needed to increase the power density of electronics and thermal mechanical systems that are envisioned for future aircraft, spacecraft and surface ships. Typically, heat exchanger cores must increase in size in order to more effectively dissipate any increased heat loads. This is impossible in many cases, thus new concepts for heat exchanger cores/systems are required. Furthermore, any new concept/material must be competitively priced in order to make it a feasible alternative for system implementation[1].

Concerning thermal management approaches, many techniques (e.g., micro-channels and heat pipes) have been investigated in order to improve efficiencies. One approach that has been employed uses metallic foams for heat exchangers. Due to increased surface area for heat transfer, metal foams have been successfully utilized for airborne equipment, heat sinks for power electronics, heat shields, as well as air-cooled condenser towers and regenerators[2]. The limitations of these materials is based on the fact that they are porous structures of the parent materials, hence the bulk thermal conductivity of the metallic foam can be an order of magnitude less than the dense parent material. In order for foams to have reasonable bulk properties, higher ligament (i.e., the solid material network within the foam) thermal conductivities are needed. Graphitic carbon foams are excellent candidates for thermal management applications due to their extraordinarily high ligament properties[4,5]. These graphite foams have reported ligament conductivities greater than 1800 W/m·K and bulk values up to 245 W/m·K (as a comparison, fully dense aluminum = 180 W/m·K). References 5 and 6 provide an extensive review of graphite carbon foam development, prototype applications, as well as potential future uses. The following table compares the properties of typical fully dense metals, aluminum foam, and graphite foam products. Very clear advantages for the graphitic foams include the density and bulk thermal conductivities, while the low mechanical strengths are an obvious detriment.

Table 1. Property Comparison of Typical Heat Exchanger Materials

Material	Type	Density (g/cc)	Bulk Thermal Conductivity (W/m·K)**	Compressive Strength (MPa)**	Porosity (%)	Open Porosity (% of total)
Aluminum	Solid	2.7	180	137	-	-
Titanium	Solid	4.5	22	450	-	-
Invar®	Solid	8.0	13	725	-	-
ERG Aerospace Duocel® Aluminum Foam*	Foam	0.2	6	3	92	-
POCO Foam*	Foam	0.5	135	3	75	96
POCO HTC*	Foam	0.9	245	6	61	95
Touchstone CFoam25®*	Foam	0.4	25	15	82	-
GrafTech***	Foam	0.15-0.60	31-175	0.2-1	70-93	

*Foam data obtained from ERG, POCO, Touchstone websites
**Out-of-plane for foams.
***Under development.

DURABILITY AND PERFORMANCE REQUIREMENTS
For graphite foams to be fully utilized for new solutions, numerous issues need to be studied and resolved. Improvements need to be made regarding the overall foam strength; flaking and pressure drop in forced convection systems; bonding; environmental durability; as well as cost, product size, quality and production capacity. Note that, currently, the production of foams is limited to small quantities with varying quality and a rather high unit cost. This production concern is the focus of numerous companies and government lead projects[1]. Furthermore, to allow for revolutionary steps to be made in thermal management systems, the graphite foams need to be part of a systematic solution that incorporates other game changing materials such as carbon/carbon composites as well as graphite heat spreaders. The complete system should aim to maximize the thermal efficiency, reduce weight and volume, increase the mean time to failure, as well as be cost effective.

Adding to the above needs for the implementation of the graphite foam based thermal management approach, basic design parameters and modeling approaches are required and a blank sheet mentality adopted. This includes appreciating and tackling issues related to system behavior when foams are combined with other advanced heat sink/spreading technologies (e.g., anisotropic graphite composite plates; phase change materials in thermal energy storage devices). As is the case with any new and advanced material development, extensive experimentation and modeling at both the material and system levels are required prior to acceptance by the design community.

OBJECTIVES

The overall goals of this effort, initiated in the fall of 2008, are to develop the production capabilities and supporting technologies needed to design and fabricate advanced graphite foam based heat exchangers and/or thermal energy storage devices for aerospace systems. An early assessment of various military aerospace systems in need of thermal management improvements found that directed energy weapons components warranted the most attention at this time (Figure 1). Directed Energy Weapons (DEW: e.g., high energy tactical laser and airborne active denial high power microwave) applications face several challenges due to the inherent inefficiencies of the systems (for some systems 90% of the input energy is wasted as heat)[8-11]. Airborne DEW call for effective and light weight thermal management for acquisition of heat from the power source, temporary energy storage, and ultimately rejection to the environment. Such a system requires high heat flux, fast response thermal energy storage devices, and effective air cooled heat exchangers for heat rejection into an environment with low thermal potentials.

Figure 1 Future high energy laser system[9].

To achieve the above stated goals, the team is further enhancing the foam manufacturing process; utilizing coatings for enhanced foam durability and performance; developing material level models for describing the basic properties as a function of foam characteristics (pore size, solid volume fraction, ligament thermal conductivity, bulk density, etc.); conducting extensive material characterization studies (thermal, mechanical, flow/pressure drop, and etc.); developing and optimizing bonding/brazing strategies for enhanced integration with face sheets and heat exchanger housing components; addressing nondestructive evaluation approaches for quality assessment and characterizing damage during service conditions; designing/optimizing multiple levels of heat exchanger systems using finite element analysis and computational fluid dynamics; and finally, producing a preliminary prototype heat exchanger that utilizes the graphite foam.

Foam Production

In all current production methods[5,6], graphite foams begin with hydrocarbon raw material because, among other things, of its high-carbon yield and graphitizability. The precursor is heated at a controlled rate in a non-oxidizing environment under pressure. The precursor melts and, as the temperature is increased, lower molecular weight components distill and volatilize. Thermal cracking occurs at higher temperatures generating additional volatiles and gases. The volatiles and gases produce bubbles that rise, creating shear forces that tend to orient the aromatic molecules. Eventually, with continued heat treatment, the viscosity increases to the point whereby the foam is transformed into a rigid, infusible structure. Additional heating is required to drive off residual volatiles and to consolidate the foam, followed by graphitizing at temperatures above 2,500°C to develop crystallinity and to promote thermal conduction.

GrafIHX™ graphite foams are being produced by GTIH with a range of densities and porosities by varying raw materials and processing conditions. Generally, the density of the graphite foams are kept below 0.6 g/cc. Figure 2 shows the typical open cell structure and the graphitic nature of the foams. Figure 3 displays the progress made thus far regarding billet sizes. The fabrication process is very repeatable as witnessed by the consistent density throughout the large billets. Typically, properties in the rise direction are superior to the other directions. Preliminary tests show conductivities to be a factor of two higher in the rise direction while mechanical strengths are approximately 10% superior.

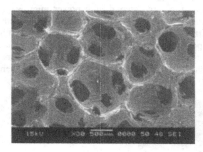 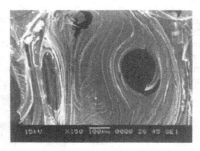

Figure 2 SEM of graphite foams showing cell morphology and graphitic structure[7].

Figure 3 Improved production capacities showing increasing billet size. Within first year, billet size grew from cubes with 15 cm sides to 50 cm diameter x 25 cm high cylinders.

Mechanical and Transport Behavior

Because of graphite foam's recent debut (1990s), extensive and detailed mechanical data is limited. Also, due to the unique morphology and fragile nature new experimental procedures for compressive and tensile testing are being designed, tested and verified. A laser extensometer and a 3D image correlation method are available for strain assessment. Figure 4 shows some preliminary stress/strain plots for compression and tension for multiple densities. As indicated in the figures, the deformation behavior is nonlinear and rather complex. Additional data indicates a highly stochastic mechanical response (deformation and strength), thereby, calling for probabilistic approaches.

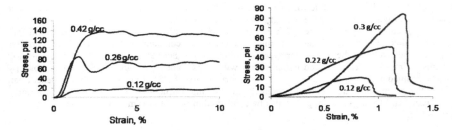

Figure 4 Typical stress/strain responses under compression (left) and tension (right) for GTIH graphite foams of various densities.

In addition to modeling moduli and strengths empirically as a function of density (e.g., Gibson-Ashby proportionality relationships[12]), finite element models are being derived from 3D x-ray computed tomography data sets (see Figure 5). The models will also be instrumental in assessing the transport behaviors and developing fluid flow models. The complex 3D models will be used to evaluate the influence of foam morphology on overall deformation response and strength as well as defining the effects of microstructure on fluid flow. Accurate models will assist in the selection of optimized foams for specific applications. Lastly, extensive experimentation is planned for studying thermal diffusivity, conductivity, permeability and etc. by employing standard test procedures (e.g., laser flash, guarded hot plate) as well as utilizing customized approaches (e.g., tailored wind tunnel experiments, flash thermography for large piece diffusivity measurements).

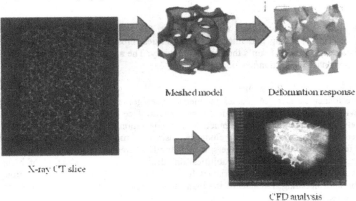

Figure 5: Finite element analysis results for models created from x-ray computed tomography data.

Bonding, Integration, and Coatings
To fully realize the many attributes of graphite foams, bonding techniques must be optimized. This includes using bonding agents that fully wet and react to all members of the structure, provide the least thermal resistance, and give suitable strength so as not to initiate mechanical failure. This project

will study the joining of graphite foams to metallic systems (Ti, Inconel 625, Cu-clad Mo, and other metals) using high conductivity thermal interface (braze) materials.

Two commercial active metal brazes (Cusil-ABA and Palcusil-5) in two forms, paste and foil, will be studied (Figure 6). These compounds have thermal conductivities of 180 W/mK and 208 W/mK, respectively, so the braze interlayer will minimize the thermal barrier in brazed heat rejection systems. Joint microstructures and compositions will be examined by scanning electron microscopy (SEM) coupled with energy dispersive X-ray spectrometry (EDS). Evidence of braze penetration within the given substrates will be presented through microstructural analysis and microhardness measurements. Mechanical testing will also be conducted in order to examine the physical strength of the joint region.

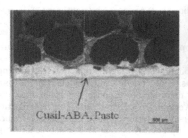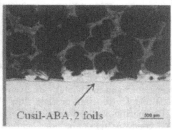

Figure 6 Optical photomicrograph showing good braze joint between graphite foam and titanium plate, most notably when using paste.

Silicon Carbide (SiC) coatings are also being developed for improving environmental durability and strength, bonding and integration capabilities, erosion and flaking resistance, while aiming to maintain the graphite foam's thermal conductivity. The approach taken will involve Silicon Oxide (SiO) vapor phase infiltration/reaction.

CONCLUSIONS AND FUTURE WORK

Graphite foams are ideal candidates for thermal management systems due their high specific bulk conductivity, ultrahigh ligament conductivity, large surface area, as well as the potential for elevated temperature applications. Prior to practical implementation in thermal management devices, a few issues need to be addressed. These include increasing the mechanical strength, controlling the pressure drop for air or fluid heat exchangers, optimizing bonding and integration strategies, as well developing large scale manufacturing capabilities aimed at continuous production of high quality, affordable billets. In parallel, the material will need to be fully characterized through both experimentation and modeling. Because of the many unique features, new approaches to testing and analysis are required.

As the project progresses, the team will design and fabricate a prototype thermal management device based on feedback from the Air Force and industry. This will provide the opportunity for showcasing the graphite foam and the associated technologies, whereby allowing for a direct comparison with contemporary designs (e.g., metallic finned structures).

ACKNOWLEDGEMENTS

The authors would like to thank Dr. Ajit Roy (Technical Monitor and Sr. Material Research Engineer at AFRL) and Dr. Alan Janiszewski (Vice President for Materials & Manufacturing at Universal Technologies Corp.) for their guidance and funding of this project under U.S. Air Force contract FA8650-05-D-5807. The authors would also like to acknowledge contributions from team members Dr. Helen Mayer from GrafTech International and Project Manager Ms. Patricia Grospiron from OAI.

REFERENCES

[1] R. Watts, G. Maxwell, M. Steenwyk, J. Biels, S. Chang, and T. Carroll, "Updates for Transition Potential of Light Weight Composite Aircraft & Spacecraft Thermal Management Components," *SAMPE Fall Technical Conference and Exhibition*, Cincinnati, Ohio, Oct. 29-Nov. 1 (2007).

[2] M.F. Ashby, et al., "Metal Foams: A Design Guide," Butterworth-Heinemann, Burlington, Maine (2000).

[3] M.D. Sarzynski, "Carbon Foam Characterization: Sandwich Flexure, Tensile and Shear Response," Thesis, Texas A&M University, December (2003).

[4] J. Klett, R. Hardy, E. Romine, C. Walls, T. Burchell, "High-Thermal-Conductivity, Mesophase-Pitch-Derived Carbon Foams: Effect of Precursor on Structure and Properties," *Carbon*, **38**(7), pp. 953-973 (2000).

[5] M. Sheffler and P. Colombo, "Cellular Ceramics: Structure, Manufacturing, Properties and Applications," Wiley-VCH Verlag GmbH & Co., Weinheim (2005).

[6] J. Klett, "High Thermal Conductivity Graphite Foam," ORNL home page for graphite foam research, http://www.ms.ornl.gov/researchgroups/CMT/FOAM/foams.htm (2010).

[7] P. Stansberry, E. Pancost, Y. Xiong, and J. Norley, "Graphite Foam Performance in Heat Exchanger Applications," SAMPE Fall Technical Conference, Wichita, Kansas, October 19-22 (2009).

[8] S. Adams and J. Nairus, "Energy Conversion Issues for Airborne Directed Energy Weapons," 37th Intersociety Conversion Engineering Conference, IECEC 2002 20047 (2002).

[9] S. Mueller, "Unique Technologies Update/Directed Energy," 34th Air Armament Symposium (AAC), Eglin Air Force Base, Florida (2008).

[10] D.L. Vrable and B.D. Donovan, "Thermal Management for High Power Microwave Sources," 1st International Energy Conversion Engineering Conference, Portsmouth, Virginia, August 17-21 (2003).

[11] G. Warwick, "Solid-State Laser Programs Advance," Aviation Week, Jan 11 (2009).

[12] L.J. Gibson and M.F. Ashby, "Cellular Solids, Structure and Properties," 2nd Ed., Cambridge University Press, Cambridge, UK (1997).

INTEGRATION OF GRAPHITE FOAMS TO TITANIUM FOR THERMAL MANAGEMENT APPLICATIONS

M. Singh[1], Rajiv Asthana[2], C.E. Smith[1], A.L. Gyekenyesi[1]

[1]Ohio Aerospace Institute, 22800 Cedar Point Road, Cleveland, OH 44142 USA
[2]Engineering & Technology Department, University of Wisconsin-Stout, Menomonie, WI 54751 USA

ABSTRACT
    The high-conductivity graphite foams are attractive for thermal management applications in avionic heat sinks and heat exchangers. However, integrating foams in thermal systems requires robust and thermally conductive joints between foam and metallic constituents. In this study, graphite foams with different densities were vacuum brazed to titanium using the active braze alloys, Cusil-ABA and Palcusil-5 (in paste and foil forms). The joint microstructure and elemental composition were examined using optical microscopy (OM) and scanning electron microscopy (SEM) coupled with energy dispersive spectroscopy (EDS) to evaluate joint integrity, interface microstructure, and chemical interaction. The low-density foams exhibited braze penetration with penetration distance increasing with decreasing bulk density. The carbon/braze interfaces were enriched in Ti indicating sound chemical bonding. The overall thermal resistance of foam/Ti joints was estimated using 1-D, steady-state heat conduction analyses for planar geometries. These calculations indicate a marginal effect of braze saturated foam on joint conductivity for practical range of process parameters.

INTRODUCTION
    In recent years, due to increased performance in a wide range of engineered products ranging from computer processors to advanced aerospace vehicles, efficient transport and removal of heat has become quite critical. Effective thermal management in these systems demands high thermal conductivity materials with low coefficients of thermal expansion (CTE) for thermal stability and for compatibility with other materials. Innovative component design and process development for effective thermal management is often coupled to the development of new and more efficient materials and technologies.
    In the repertoire of thermal management materials [1-10] are conductive metals such as Al and Cu, clad metals such as Cu-clad-Invar and Cu-clad-Mo, composites such as C-C, Cu/W, SiC/Al, C/Cu, C/Al, B/Al, and BeO/Be, and porous ceramics. Recently, porous ceramics and porous graphite have attracted considerable interest in thermal management applications [11-13]. Porous materials can be designed to maximize either the thermal resistance or heat transfer with a flowing fluid at considerable weight savings. Thus, thermal management using porous materials could involve complex interaction between several different factors.
    Graphite foams of variable densities are excellent materials for applications in heat sinks and heat exchangers due to their high relative thermal conductivity (out-of-plane conductivity: 135-245 W/m-K, in-plane conductivity: 45-70 W/m-K). In these foams, the combination of open pore structure, dense graphitic matrix, and high conductivity is beneficial to high heat transfer effectiveness. Integration of planar and tubular titanium subcomponents and graphite foam could potentially enhance the heat transfer in light-weight avionic heat exchangers. Recently, integration of high-conductivity C/C composites with Ti, Inconel 625, and controlled expansion Cu-clad-Mo alloys was demonstrated for increased functionality at reduced weight in thermal management applications. Active metal brazing was used for joining [14-18]. While carbon and graphitic materials have been soldered for low use temperatures, new challenges and issues become important when such joints are created for high temperature applications.

In this study, integration of various commercially available graphitic foams with different bulk densities (varying porosity content) to commercially pure Ti using active metal brazing was demonstrated. The joint microstructure and composition were characterized using optical microscopy, scanning electron microscopy and energy dispersive spectroscopy. As the foam/metal joints are being developed for thermal rather than structural applications, the thermal resistance of the joint is also critically important along with the bond strength. The thermal resistance of foam/Ti joints was estimated for planar geometries to assess the effects of foam and metal substrate conductivities, depth of braze penetration, and metal-to-foam thickness ratio on the overall thermal resistance of joined assemblies.

EXPERIMENTAL PROCEDURE

The porous graphitic foams used in joining were obtained from GrafTech International Holdings (GTIH), Parma, OH and Poco Graphite, Inc., Decatur, TX. Some typical physical properties of these foams are listed in Table 1.

Table 1: Typical properties of graphitic foams used in present study.

Foam Supplier/Foam Type	GrafTech Foam*	POCO Foam
Bulk density (g/cc)	0.15-0.60	0.50-0.90
Solid volume fraction (%)	7-30	25-39
Pores/inch (ppi)	20-40	64
Avg. Pore size (µm)	1300-600	400
Bulk Conductivity (z-axis) (W/m.K)	100-175	135-245
Ligament Conductivity (W/m.K)	500-700	540-630

* Foams under development.

Two commercial active metal braze alloys, Cusil-ABA and Palcusil-5 in powder as well as foil form were obtained from Morgan Advanced Ceramics, Hayward, CA. The composition, liquidus and solidus temperatures, and selected physical and mechanical properties of the braze alloys are given in Table II.

Table 2. Selected Properties of the braze alloys and Titanium used for brazing

Braze	Braze Composition (wt%)	T_L, C	E, GPa	YS, MPa	UTS, MPa	CTE×10^6, K^{-1}	% El	K, W/mK
Palcusil-5®	68Ag-27Cu-5Pd	810	-	333	380	17.2	11	208
Cusil-ABA®	63Ag-35.3Cu-1.75Ti	815	83	271	346	18.5	20	180
Ti	CP-Grade 2		105	480	550	8.6	15	17.2

E: Young's modulus, YS: yield strength, TS: tensile strength, CTE: coefficient of thermal expansion, %El: percent elongation, K: thermal conductivity. ® Morgan Advanced Ceramics, Hayward, CA.

Commercially pure Ti plates obtained from Timetal, Inc., were sliced into 2.54 cm x 1.25 cm x 0.25 cm pieces. All materials were ultrasonically cleaned in acetone for 15 min. prior to brazing. The braze foils were sandwiched between the metal and the graphite foam, and a normal load of 0.30-0.40 N was applied to the assembly. Braze foils are easier to use than braze powders especially for small gaps in which powder paste application could be difficult. Additionally, the residual organic solvents in powder pastes could cause soot formation and furnace fouling. However, as braze powders are used in industrial work, braze runs were made using braze powders in place of foil in order to examine the differences, if any, when using foils and powders. For this purpose, braze powders were mixed with glycerin to make a thick paste with dough-like consistency, and the paste was applied using spatula to the surfaces to be joined. The assembly was heated in an atmosphere-controlled furnace to the brazing temperature (typically 15-20 C above the braze liquidus) under vacuum (10^{-6} - 10^{-5} torr), isothermally held for 5 min. at the brazing temperature, and slowly cooled to room temperature.

The joined samples were visually examined, then mounted in epoxy, ground and polished on a Buehler automatic polishing machine using the standard procedure, and examined using optical microscopy (Olympus DP 71 system) and scanning electron microscopy (SEM) (JEOL, JSM-840A) coupled with energy dispersive x-ray spectroscopy (EDS). The elemental composition across joints was accessed with the EDS and presented as relative atomic percentage among the alloying elements at point markers on SEM images.

RESULTS AND DISCUSSION
Microstructure and Composition

Figure 1 (a) and (b) shows the interfacial microstructure of GTIH foam/titanium joints fabricated using the braze foils and pastes, respectively. The extent of braze penetration in the foam is better using the paste than the foils. The approximate depth of braze penetration (estimated from low-magnification photomicrographs) for all foams was typically 1.5-4.0 mm. However, there was only surface reaction (no substantial infiltration) in high-density foam.

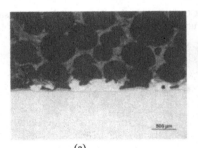 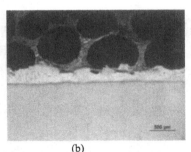

(a) (b)

Fig. 1: Micrographs of foam/Cusil-ABA/Titanium joints (a) two Cusil-ABA foils; and (b) Cusil-ABA paste.

Representative secondary electron images of joint microstructures and elemental composition scans from EDS are shown in Fig. 2 for joints made using two GrafTech foams: Foam A3 (density ~ 0.246 g/cm^3) and Foam A10 (density ~0.481 g/cm^3). Intimate bonding is evident between Cusil-ABA and foam ligaments in joints. Titanium enrichment is noted in the interaction zone (Fig. 2(c) & (f)) between graphite and Cusil-ABA. There is no evidence of carbon dissolution in braze or Cu, Ag or Ti dissolution in carbon (Fig. 2(c)). Titanium dissolution has formed a Cu-Ti phase in the vicinity of the carbon/braze interface (Fig. 2(f)).

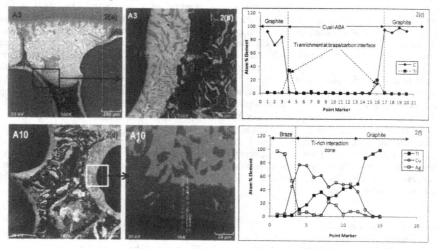

Fig. 2: Secondary electron images of braze infiltrated (a) & (b) A3 Foam (density ~ 0.246 g/cm³) and (d) & (e) A10 Foam (density ~ 0.246 g/cm³); (c) & (f) composition at point markers in (b) & (e), respectively.

The optical micrographs of GrafTech and POCO foams brazed to Ti using Palcuil-5 are shown in Figs 3(a) and (b). The joint microstructures appear uniform with well bonded foam ligaments.

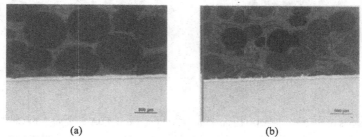

(a)   (b)

Fig. 3: Optical photomicrograph of (a) GTIH foam/Palcusil-5/Ti; and (b) POCO/Palcusil-5/Ti joints fabricated using two braze alloy foils.

The major braze constituents Ag and Cu do not wet graphite and display large values of contact angle θ (~137°-140°); however, Ti additions to Ag and Cu improve the wettability because of the reaction-induced wetting by carbide forming reactions. Usually, small quantities of Ti and short brazing times suffice for maximum spread. Thus, Ti segregation at the carbon/Cusil-ABA interface in

Fig. 2 is consistent with the wettability enhancement from the chemical modification of the interface. The contact angle of Palcusil-5 on carbon could not be found; however, carbon supported Pd catalysts are widely used, and formation of carbides (e.g., PdC, metastable Pd_6C, sub-stoichiometric $PdC_{0.15}$ etc.) is well established. Carbon is also known to dissolve interstitially in Pd, forming a FCC solid solution. Thus, dissolution and carbide formation shall promote the wettability of Palcusil-5 on the graphite foam.

Interfacial Thermal Resistance

For our joints, the thermal property data of various materials are summarized in Tables 1 and 2. For the thermal resistance calculations, a planar composite wall consisting of slabs of Ti and foam with an intervening layer of braze is modeled as shown in Fig. 4(a).

For 1-D steady-state heat conduction, the joined materials form a series thermal circuit with an effective thermal resistance, $R_{eff} = \Sigma(\Delta x_i/K_i)$, where Δx_i and K_i are the thickness and thermal conductivity, respectively, of the ith layer. Because the braze is likely to penetrate the foam to some distance, the system essentially is comprised of Ti, braze layer, the infiltrated foam, and un-infiltrated foam. The effective thermal resistance can be written as:

$$R_{eff} = (\frac{\Delta x_{Ti}}{K_{Ti} A}) + (\frac{\Delta x_{braze}}{K_{braze} A}) + (\frac{\Delta x_{inf}}{K_{eff} A}) + (\frac{\Delta x_{foam}}{K_{foam} A})$$

Where A is the cross-sectional area ($A = 6.45 \times 10^{-4}$ m^2). A mass balance for total braze material including the infiltrated foam (thickness: Δx_{inf}) and the residual braze in the joint region (thickness: Δx_{braze}) yields: $(1 - \varphi_{gr})\Delta x_{inf} + \Delta x_{braze} = X_{braze}$. φ_{gr} is the volume fraction of graphite in the foam, and X_{braze} is the known initial braze thickness in the joint prior to processing. Based on experimental observations, Δx_{inf} was assessed to be 25μm. The mass balance equation was then utilized to solve for Δx_{braze}.

The results presented in Fig. 4(b-d) reveal only marginal changes in thermal resistance of the joined assembly for depths of penetration observed in actual experiments. A more realistic situation would require one to more accurately account for the thermal resistance of the foam that is partially filled with braze. Indeed, visual observations revealed that braze flow had occurred over large distances within the porous foam but without complete pore filling and closure. Thus, a large proportion of ligaments (struts) become coated with braze but the pore volume only marginally decreases.

The thermal conductivity of foam is comprised of four contributions: conduction through solid, conduction through gas, convection within cells, and radiation through cell walls and across voids. The braze conductivity (180 W/m.K) is significantly less than ligament conductivity in POCO foam (540-630 W/m.K) and GrafTech foam (460-700 W/m.K). Also, considering the conductivity of air at 1 atm pressure to be 0.0263 W/m.K at 300 K [19], and its volume fraction in un-infiltrated POCO foam to be 61 to 75%, the gas conductivity within the foam should then be 0.016-0.0197 W/mK at 300 K. Thus, for high-conductivity graphite foams, the largest contribution to thermal conductivity is via ligaments. Conduction through air, convection within cells, and radiation from struts shall, therefore, make negligible contribution relative to conduction via ligaments. Pore closure due to braze infiltration shall actually replace conduction through gas with conduction through the impregnated braze. Braze penetration of pores by replacement of air shall thus aid thermal conduction even at the expense of increased weight. Thermal conduction via ligaments shall, however, decrease if ligaments get coated with braze and complete pore filling and closure by braze saturation do not occur. Because recipes designed to limit braze permeation into graphite foam (e.g., use of hydrophobic coatings) shall also likely be detrimental to joint formation, the best approach to maximize the thermal performance of joined assemblies would be to use the minimum quantity of braze needed to create a sound joint. This will leave little excess braze that could permeate the foam and impair the conductivity by forming a coating on ligaments.

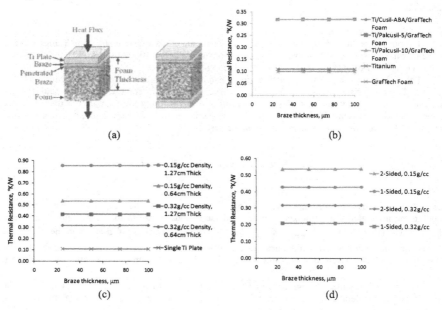

Fig. 4 (a) Schematic of the one and two-sided joint assemblies used for thermal calculations, (b) thermal resistance vs. braze layer thickness for a two-sided joint of Ti and GrafTech foam (0.32g/cc), foam dimensions are 0.64cm thickness and 2.54cm x 2.54cm cross-sectional area, and Ti plate thickness is 0.16cm (thermal resistance of the GrafTech foam and an individual Ti plate is shown for comparison); (c) thermal resistance vs. braze thickness for a two-sided joint with Ti and GrafTech foam (braze: Cusil-ABA). Joints with low density (0.15g/cc) and medium density (0.32g/cc) GrafTech foam are compared, along with 0.64cm and 1.27cm thick foam; (d) comparison of thermal resistance vs. braze thickness for 1 and 2-sided joints consisting of Ti plates, Cusil-ABA, and 0.64cm thick Graftech foam. Two different foam densities are considered (cross-sectional area is 2.54cm x 2.54cm).

SUMMARY AND CONCLUSIONS

Successful joining of commercial graphitic foams of different bulk density to pure Ti using Ti and Pd-containing brazes, Cusil-ABA and Palcusil-5, was demonstrated. The joint microstructures revealed defect-free and well-bonded interfaces, Ti segregation at the carbon/braze interface, and dissolution of Ti substrate into molten braze. Theoretical projections of overall thermal resistance of joined assemblies point toward the benefits of integrating high-conductivity graphitic foam and titanium for light-weight heat rejection systems. Rigorous thermal calculations and experimental measurements are currently underway for more definitive outcomes concerning the thermal behavior of Ti/foam joints.

ACKNOWLEDGMENTS

This work was funded through U.S. Air Force Contract FA8650-05-D-5807, Prime Contractor: Universal Technology Corporation (UTC). The authors would like to thank Dr. Ajit K. Roy for technical suggestions. We would also like to thank Dr. Alan Janiszewski, Program Manager/Vice President for Materials & Manufacturing, UTC and other project partners.

REFERENCES

1. C. Zweben, Advances in high-performance thermal management materials, *J. Adv. Mater.*, 2007, 39(1), 3-10.
2. J. W. McCoy and D. L. Vrable, Metal-matrix composites from graphitic foams and copper, *SAMPE Journal*, 40(1), 2004, Jan/Feb 2004, 7.
3. E. Neubauer, P. Angerer, and G. Korb, Heat Sink Materials with Tailored Properties for Thermal Management, *28th Int. Spring Seminar on Electronics Technology, 2005 IEEE*, 258-263.
4. J. Norley, J.J.-W. Tzeng, G. Getz, J. Klug and B. Fedor, The Development of a Natural Graphite Heat-Spreader, *17th IEEE SEMI-THERM Symposium, 2001, IEEE*, 107-110.
5. T. W. Clyne, I. O. Golosnoy, J. C. Tan and A. E. Markaki, Porous materials for thermal management under extreme conditions, *Phil. Trans. R. Soc.*, A (2006) 364, 125–146
6. R. Watts, G. Maxwell, J. Arnold, B. Pesta, Transition Potential of Light Weight Composite Aircraft & Spacecraft Thermal Management Components, *SAMPE 2006 Fall*, Society of Advanced Materials and Process Engineering, CA, 2006
7. M. Kistner, R. Watts, and A. Colleary, Materials Opportunity to Electronic Composite Enclosures for Spacecraft and Aerospace Thermal Management, in Proceedings of *Space Technology and Applications International Form (STAIF-06)*, M.S. El-Genk (ed.), American Institute of Physics, Melville, New York, 2006
8. J. Banisaukas, M. Shioleno, C. Levan, R. Rawal, E. Silverman, R. Watts, Carbon Fiber Composites for Spacecraft Thermal Management Opportunities., in Proceedings of *Space Technology and Applications International Form (STAIF-05)*, M.S. El-Genk (ed.), American Institute of Physics, Melville, New York, 2006
9. F. Beaver, D. Vrable, R. Watts, Application of Carbon-Carbon Heat Exchanger for Aircraft, *Paper # 981291, SAE*, 400 Commonwealth Drive, Warrendale, PA, 15096-0001, 1998
10. S.A. McKeown, R.D. LeVasseur, 'High performance heat sink for surface mount applications', *CH 3030-4/91/0000-0153, IEEE*, 1991, 153-157.
11. L.J. Gibson and M.F. Ashby, *Cellular Solids*, 2nd ed., Cambridge University Press, 2001, 285-289.
12 . J.W. Klett, Mesophase Pitch-Based Carbon Foam: Effects of Precursor on Thermal Conductivity, Proceedings of the *23rd Annual Conference on Ceramic, Metal and Carbon Composites, Materials, and Structures*, Jan. 25-28, 1999, Cocoa Bch, FL, 657-674.
13. POCO HTC web site: http://www.poco.com/us/Thermal/htc.asp#thermal (August 22, 2006)
14. M. Singh and R. Asthana, Characterization of brazed joints of carbon-carbon composites to Cu-clad-Mo, *Compos. Sci. Tech.*, 68, 14 (2008) 3010-3019
15. M. Singh, G.N. Morscher, T.P. Shpargel and R. Asthana, Active metal brazing of titanium to high-conductivity carbon-based sandwich structures, *Mater. Sci. Eng. A*, 498, 1-2, 2008, 31-36.
16. M. Singh, R. Asthana and T.P. Shpargel, Brazing of C-C composites to Cu-clad Mo for thermal management applications, *Mater. Sci. Eng. A*, vols. 452-453, 2007, 699-704.

17. G.N. Morscher, M. Singh, T.P. Shpargel and R. Asthana, A simple test to determine the effectiveness of different braze compositions for joining Ti tubes to C/C composite plates, *Mater. Sci. Eng. A*, 418(1-2), 2006, 19-24.
18. M. Singh, T.P. Shpargel, G.N. Morscher and R. Asthana, Active metal brazing and characterization of brazed joints in titanium to carbon-carbon composites, *Mater. Sci. Eng. A*, 412, 2005, 123-128.
19. K.D. Hagen, *Heat Transfer with Applications*, Prentice Hall, 1999

FABRICATION OF NOVEL HEAT INSULATOR USING POROUS CERAMICS MATERIALS

Kazuma Kugimiya, Mitsue Ogawa and Hideaki Matsubara
Japan Fine Ceramics Center (JFCC)
2-4-1 Mutsuno, Atsuta-ku, Nagoya, 456-8587 Japan

ABSTRACT

Porous silica powder and transparent porous silica were synthesized for the development of thermal insulating materials and their thermal conductivity was evaluated. It was clarified that the thermal conductivity of the porous ceramic material was lowered than that of simple vacuum at lower vacuum region (higher pressure), indicating the porous ceramic materials had an advantage for the application of thermal insulator.

1. INTRODUCTION

A lot of energy was consumed by air conditioning for houses and buildings, the energy consumption of which makes up 8% of the total in Japan. In order to reduce the amount of energy consumption, improving the thermal insulation performance of houses, buildings, etc is the most effective way.

The heat insulating material used in houses and buildings (walls, floors, etc.) requires not only a high thermal insulation property but also excellent compression strength. Moreover, incombustible and lightweight material is favored for thermal insulation material. In addition to energy conservation within houses and buildings, development of effective heat insulating materials for home appliances, transportation vehicles, energy storage, etc. is also required. It would be meaningful if we developed effective heat insulating wall materials and window materials by utilizing high-strength (compression) heat insulating ceramic particle technology, ceramic/polymer compound technology, high-efficiency radiation prevention coating technology, transparent high-performance heat insulating material technology to contribute to energy conservation and CO_2 reduction efforts. In order to realize wall materials and window materials that exhibit much improved thermal insulation performance and bring about a considerable energy-saving effect in air conditioning of houses and buildings, home appliances, transportation equipment and energy storage, we will develop multi-ceramics that suppress any type of heat transfer, such as lattice vibration, convection, and radiation, and also develop technology for compounding polymer and glass.

In this paper, porous silica powder and transparent porous silica were synthesized for the development of thermal insulating materials. In order to clarify an advantage of porous ceramic material comparing with simple vacuum condition for the application of vacuum thermal insulator panel, we measured the pressure dependence of thermal conductivity for porous ceramics materials such as powder and silica aerogel.

2. EXPERIMENTAL

2.1 Synthesis of porous silica powder

In order to develop multi-ceramic materials that suppresses any of the three elements of heat transfer (lattice vibration, convection and radiation), it is necessary to develop technology for synthesizing ceramic particles possessed of a nano-porous structure. We synthesized porous silica powder prepared by precipitation of the colloid-state silica precursor and subsequent drying process. The liquid glass was used as a silica source.

2.2 Synthesis of transparent porous silica

The silica aerogels, transparent ceramics materials, were prepared by sol-gel process and subsequent supercritical CO_2 method. In our research, Metylsilicate 51(Colcoat Co. Ltd.), prepared by condensation tetrametyl orthosilicate to oligomerize it into the average tetramer, was used as the silica source. T.M. Tillotson et al reports that silica aerogels prepared by using two stage approach is superior to single-step base catalyzed aerogel in optical and mechanical properties[1]. For two stage approach, the first step is to form a partially hydrolyzed, partially condensed silica precursor by reacting tetra-alkoxysilane with a sub-stoichiometric amount of water. This precursor is then further processed to a gel in the second step which completes the hydrolysis under basic conditions. Metylsilicate 51 possesses the similar structure as the silica precursor at the first step, therefore, the aerogels prepared by Methylsilicate 51 is expected to exhibit the excellent properties comparable to two step aerogels. Moreover, Metylsilicate 51 is commercially available at a moderate price, so we adopted the Metylsilicate 51 as silica source of aerogel.

Silica alcogel was prepared by mixing Metylsilicate 51, ethanol, 0.1N NH_4OH at a molar ratio of 1:40:24. After gelation, the alcogel was aged at 50 °C for 36 hours. In order to metylate alcogels, the gels were immersed into 2 mol/L 1,1,1,3,3,3 hexamethyldisilazane at room temperature for 48 hours. Metylated alcogel was washed in ethanol three times every twelve hours. Finally, in the autoclave (SCRD6,

Ryusyo Industrial co., Ltd.), the alcogels were dried by supercritical carbon dioxide at 50 °C, 12 MPa with a flow rate of 20mL/min.

2.3 Measurement of thermal conductivity and other properties

In order to probe the microstructure of the porous ceramics materials, Field Emission-Scanning electron microscopy (FE-SEM) observation (SU8000, HITACHI) was carried out without coating metal at a low accelerating voltage. The porous ceramics powder was subjected to grinding by planetary ball mill for 10 minutes. For silica aerogel, the roughly crushed specimen was performed by SEM observation.

The textual properties of porous ceramics materials such as BET surface area, pore volume and average pore diameter were investigated by a surface area analyzer (Quantachrome, AUTOSORB-1). The specimens were initially degassed at 150°C for 2 hours to remove the adsorbed species and the N_2 adsorption isotherms were obtained at -196°C.

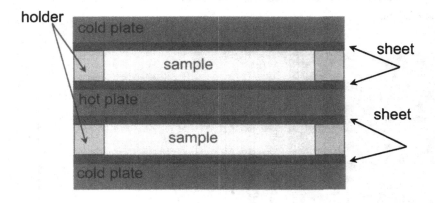

Figure 1. Schematic drawing of Guarded Hot Plate method in order to measure the thermal conductivity of porous ceramic powder.

The measurements of thermal conductivity were carried out using a guarded hot plate method apparatus (GHP456 Titan, NETZSCH)[2]. Figure 1 illustrates a schematic drawing of guarded hot plate method to measure the gas pressure dependence of thermal conductivity[3, 4] for porous ceramic powders and silica aerogels fabricated in this study. The measurement specimen, which is measured their thickness in advance, was placed above and below the hot plate and sandwiched between cold plates temperature controlled by coolant and heater. In equilibrium, thermal conductivity, λ, was defined as follows;

$$\lambda = -\frac{0.5 \cdot \phi \cdot d}{A \cdot \Delta T} \tag{1},$$

where ϕ is the heat flow through the meter area of the specimen, d is thickness of specimens, A is meter area normal to direction of heat flow, ΔT is temperature difference across specimens. In order to measure the accurate heat flow through the specimen, hot plate is surrounded by guard ring heater which maintained at the same temperature of hot plate due to suppressing horizontal escaping of heat. The temperate difference is 10K and average temperature is 25°C. The size of hot plate is 15cm by 15 cm and the system is designed for square samples 30 cm by 30 cm in size.

In order to measure the thermal conductivity at vacuum state, the powder sample is measured with the powder filled into the calcium silicate container covered both side by the black body treated aluminum foil. Inside dimension of container was 25 cm square and 1.3 cm thickness and outside dimension was 30 cm square. As for aerogel, measurement was conducted with the 9.5cm×9.5cm×1cm aerogel tile lined with sample space.

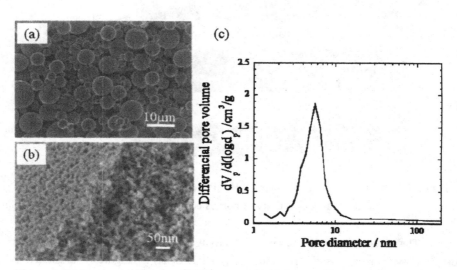

Figure 2. (a) Low magnification SEM images, (b) high magnification image and (c)pore size distribution of porous silica powder.

3. RESULT AND DISCUSSION

3.1 Structural characterization; SEM observation, Nitrogen adsorption

The low magnification SEM image for porous silica powder is shown in Figure 2(a). The particle size of silica powder was around 5 μm and the particle with the extremely large and small particle size was not observed. One can see that the surface of the particle looks smooth and the resultant particle has the well-defined spherical shape. Figure 2(b) shows the high magnification SEM images of the crushed porous silica powder. The SEM images indicated that the pore structure was observed not only at the surface of the powder but also inside the powder and the silica powder has the open pore structure with the pore diameter of approximately 10 nm. This pore size is in good agreement with the result of nitrogen adsorption, 7 nm, as shown in Figure 2(c). BET surface area of silica powder is 460 m^2/g and pore volume is 0.617 cm^3/g.

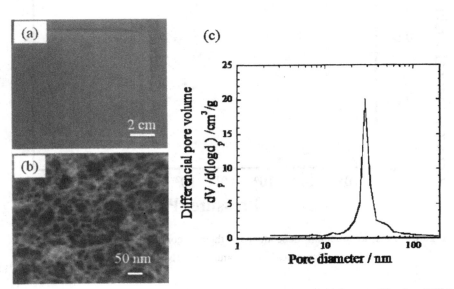

Figure 3. (a) Appearance of transparent silica aerogel. (b) High magnification SEM image of the silica aerogel. (c) pore size distribution of silica aerogel.

Figure 3(a) shows outlook of a monolithic silica aerogel. We succeeded in the supercritical drying of monolithic silica aerogel without crack. The size of the monolithic aerogel was $9.5 \times 9.5 \times 1$ cm^3 and the shrinkage ratio was less than 2 %. The bulk density of the aerogel was 0.12 g/cm^3, which was calculated from their mass to volume ratios. High magnification SEM image of silica aerogel is shown in Figure 3 (b). The aerogel was composed of intricate silica network on the orders of nanometers and nano-pore of about 20 nm in diameter. Figure 3(c) shows the pore size distribution of silica aerogel. The average pore size is approximately 20 nm and sharp and narrow pore size distribution is obtained. The resultant aerogel shows high BET surface area of 665 m^2/g and high pore volume of 3.21 cm^3/g.

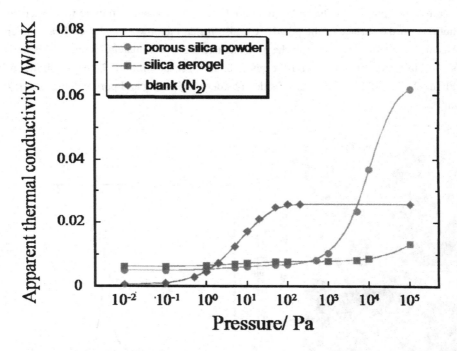

Figure 4. Pressure (vacuum) dependence of thermal conductivity for porous silica powder, silica aerogel and blank (N_2) gas.

3.2 Thermal properties

Figure 4 depicted the gas pressure (vacuum) dependence of the apparent thermal conductivity λ of porous silica powder and silica aerogel. For comparison, thermal conductivity of nitrogen gas is also shown. In high vacuum region, the thermal conductivity of nitrogen gas is almost zero, however, the thermal conductivity gradually increased with increasing the gas pressure at the pressure around 1 Pa and saturated at approximately 10^2 Pa. The value of saturated nitrogen gas conductivity was 0.026 W/m·K, which in good agreement of the reference data[5]. The behavior of pressure dependence of the thermal conductivity at the pressure from 1 Pa to 100 Pa is attributed that the mean free path of the gas molecule is comparable with the sample thickness.

The apparent thermal conductivity λ of silica powder rapidly decreased at the pressures from atmospheric to 1000 Pa, while at lower pressure than 1000 Pa, significant change was not observed. As for aerogel, apparent thermal conductivity was constant at the pressure below 10^4 Pa. The thermal conductivity of the porous powder and aerogel was lowered than that of simple vacuum at lower vacuum region (higher gas pressure). It indicates the porous ceramic materials had an advantage for the application of thermal insulator.

Apparent thermal conductivity can be expressed by;

$$\lambda = \lambda_{evac} + \lambda_{G0} / (1 + p_{1/2}/p_G), \tag{1}$$

where λ_{evac} is the thermal conductivity of the evacuated specimen (for example, gas pressure of 1 Pa), λ_{G0} is the thermal conductivity of still nitrogen gas, $p_{1/2}$ is the gas pressure, at which the gaseous conductivity is equal to $\lambda_{G0}/2$. $P_{1/2}$ is inversely proportional to the pore width Φ and, for air, is defined as follows;

$$p_{1/2} /(\text{mbar}) \approx 230 /(\Phi/\mu m). \tag{2}$$

From equation (2) and Figure 4, the pore width Φ of porous silica powder is estimated to be approximately 3 μm. The average grain size of this powder is approximately 5 μm, the grain space is expected to be a few μm. This grain space shows good accordance with the pore width between silica powders presumed by the pressure dependence of thermal conductivity. When it comes to aerogel, inadequate saturated thermal conductivity of aerogel even at an atmospheric pressure allowed us not to estimate the pore width of aerogel with high accuracy. However, according to the presumed thermal conductivity of the resultant data of thermal conductivity, the pore width is expected to be a few dozen micrometers, which is the good agreement with the pore width from the nitrogen adsorption measurement and SEM observation. This good agreement of pore width between the data from vacuum dependence of thermal conductivity and the nitrogen adsorption measurement for silica powder and aerogel indicates that the equation (2) is quantitatively-correct and the onset pressure, at which thermal conductivity increased dramatically, is higher (lower vacuum) for the substance with smaller pore size. Vacuum insulation panel is prospective to increase the thermal conductivity in the long term use due to penetrate the gas molecule through the laminate film or at the sealing faces. Therefore, it is desirable that the onset pressure at which the thermal conductivity increased is high, and then, the substance with smaller pore size is suitable for the vacuum insulation panel.

4. CONCLUSION

We synthesized porous silica powder and transparent silica aerogel and evaluated its vacuum dependence of thermal conductivity. It was clarified that the thermal conductivity of the porous ceramic material was lowered than that of simple vacuum at lower vacuum region (higher gas pressure), indicating the porous ceramic materials had an advantage for the application of thermal insulator.

Reference

[1]T. M. Tillotson and L.W. Hrubesh, J. Non-Cryst. Solids 145 (1992) 44

[2]J. Bluum, T. Denner and Y. Shinoda, Proc. Of 28th Japan Symposium on thermo physical properties (2008) 85

[3]M. Reim, G. Reichenauer, W. Korner, J. Manara, M. Arduini-Schuster, S. Korder, A. Beck and J. Fricke, J. Non-Cryst. Solids 350 (2004) 358

[4]D. Lee, P. C. Stevens, S. Q. Zeng and A. J.Hunt, J. Non-Cryst. Solids 186 (1995) 285

[5]R. A. Perkins, H. M. Roder and C. A. Nieto de Castro, J. Res. NIST 96 (1991) 247

DETECTION AND CLASSIFICATION OF GASEOUS COMPOUNDS BY SOLID ELECTROLYTE CYCLIC VOLTAMMETRY SENSORS

Grzegorz Jasinski
Faculty of Electronics, Telecommunication and Informatics,
Gdansk University of Technology, Narutowicza 11/12, 80-952 Gdansk
Gdansk, Poland

ABSTRACT

Solid state gas sensors are cheap, small, rugged and reliable. Typically electrochemical sensors based on solid electrolytes work either in potentiometric or amperometric modes. However, a lack of selectivity is sometimes a shortcoming of such sensors. It seems that improvements of selectivity can be obtained in case of the sensors based on cyclic voltammetry principle. Its working principle is based on electric current acquisition, while voltage ramp is applied to the sensor. Current-voltage response shape depends in a unique way on the type of gas and its concentration exposed to the sensor. Response of such sensors contains significantly more information comparing with typical amperometric or potentiometric sensors. Together with proper analysis method such sensors can selectively detect multiple gaseous compounds by single sensor structure.

In this study features and recent investigation results of cyclic valtammetry sensors will be presented. Most results refer to ceramic sensor based on Nasicon. Linearly changing voltage excitation signal of symmetrical triangular shape in range from 5 V to –5 V was applied to the sensors terminals. Sensor current response is recorded, which is analyzed with different methods. It is demonstrated that it can be used for the detection of mixtures of NO_2, SO_2 and synthetic air.

INTRODUCTION

In recent years, solid state ion conducting materials, also known as superionic or fast ion conductors, has been intensively developed. These materials have a relatively high ionic conductivity based on a single predominantly conducting anion or cation species and have negligibly small electronic conductivity. Typically, useful solid electrolytes exhibit ionic conductivities from 10^{-1} to 10^{-5} S/cm at room temperature. Solid state electrolytes exhibit a potential for application in a variety of solid state electrochemical devices such as fuel cells, batteries, membranes, pumps and sensors[1-3].

Gas sensors are one of the most critical and rapidly growing areas in modern solid electrolyte technology with applications in many fields, including emission control and environment monitoring. Solid state gas sensors are cheap, small, can operate on-line or under continuous deleterious conditions, rugged and reliable[4]. There is a large variety of different solid state gas sensors based on a variety of principles and materials[5]. Among them very popular become solid state electrochemical sensors, where chemical species reacting at an electronic conductor/ionic conductor interface exchange electric charges, what results in an electric signal. The output of the electrochemical sensors is directly related to the concentration or partial pressure of the gaseous species. Depending on whether the output is an electromotive force or an electrical current, the electrochemical gas sensors can be classified in potentiometric or amperometric. A lack of selectivity is usually a major shortcoming of such sensors.

It seems that improvements of selectivity can be obtained in case of the sensors based on cyclic voltammetry principle. Cyclic voltammetry is a method widely used in liquid electrochemistry for determination of chemical species concentration. The method is based on oxidation and reduction of chemical species on electrodes polarised using linearly changeable voltage. Gas concentration determination is possible using information from electrokinetic reaction occurring in a sensor. In the presence of applied voltage gases react on the surface of the electrodes, what influences current flowing through the sensor. This results in a unique voltammetric plot for different type and concentration of gases. In the case of gas sensors based on solid electrolyte instead of liquid solution the solid state electrolyte is used as ion conducting medium. As in the case of liquids electric current is measured, while voltage excitation of certain shape is applied to the sensor. Current-voltage (I-V) response shape depends in on the gas type and its concentration usually observed as current peaks in the current-voltage sensor response plot. Sometimes position of the peak can be related to the type of gas, while the peak height can be related to the concentration of this gas. Such response contains significantly more information comparing with typical amperometric or potentiometric sensors. This feature can be used for improvement of selectivity or multigas sensing. Due to the sensing mechanism (electrochemical reactions which are electrically catalyzed), such sensors are frequently called electrocatalytic sensors or sensors based on kinetic principle.

Despite its many advantages electrocatalytic gas sensors are generally little known. There are several known teams working on development of cyclic voltammetry gas sensor technology. Studies are conducted on two groups of electrocatalytic sensors: sensors working with small sinusoidal voltage excitation where the sensor works in the vicinity of thermodynamic steady state[6] and sensors working with large voltage excitation, typically triangular, with an amplitude from $\pm1V$ to $\pm5V$[7-11]. The most widely used solid state electrolytes are an oxygen ion conductors: yttria-stabilized zirconia (YSZ)[7], tungsten-stabilized bismuth oxide[8], samarium doped ceria[10], and sodium ion conductors, β"-alumina[6] and Nasicon[9]. The behavior of electrocatalytic sensors is strongly determined not only by the type of electrolyte used, but also temperature, voltage shape excitation, the rate of voltage changes and electrodes type and shape. With careful choice of sensor construction and measuring conditions desired sensors properties can be obtained. As an example for sensor based on Lisicon, change of temperature with other conditions unchanged can cause at 300°C sensor response to NO_2, at 550°C sensor response to SO_2 and at 400°C sensors response to both, NO_2 and SO_2[11].

In this study features and recent investigation results of sensors based on cyclic valtammetry will be presented. Results refer to ceramic sensor based on Nasicon and Lisicon. Properties of the developed sensor exposed to mixtures of nitrogen dioxide, sulfur dioxide and air are investigated.

SENSING PHENOMENA

In Fig. 1 the working mechanism of nitrogen dioxide electrocatalytic sensor based on Nasicon is described. The sensing phenomenon is based on formation and decomposition of a gas sensitive layer and reactivity of this layer with surrounding gases. When a voltage ramp is applied to the sensor, at the negative electrode the nitrogen dioxide reacts with sodium ions forming a layer of sodium nitrate according to the formula (eq. 1):

$$Na^+ + e^- + NO_2(g) + \tfrac{1}{2}O_2(g) \Leftrightarrow NaNO_3(s) \tag{1}$$

where 'g' and 's' indicate the gas and the solid phase, respectively. Meanwhile at the positive electrode decomposition of the already existing layer proceeds. There is another reaction possible (eq. 2), too:

$$Na^+ + e^- + NO_2(g) \Leftrightarrow NaNO_2(s) \tag{2}$$

The electric current passing through the sensor is the sum of the currents carried by the formation and decomposition of the $NaNO_3$ and/or $NaNO_2$ layers at the two electrodes. Different gas sensitive layers have different kinetics of formation and decomposition. Speed of a chemical reaction is the fastest for a specific voltage, which is characteristic for a given chemical reaction. However, the sodium ions must be supplied from the bulk to the surface of Nasicon in order to proceed the composition of the layer. The speed of sodium ion transport is connected with diffusion processes, which limit the speed of electrode reaction. As a result current peaks may be visible on current –

voltage response of the sensor. The height of the peak or the peak area can be related to the concentration of nitrogen dioxide.

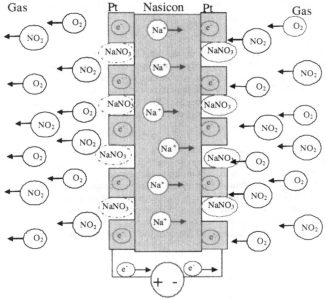

Figure 1. Sensing mechanism of nitrogen dioxide electrocatalytic sensor based on Nasicon

In the presence of more then one toxic gas in the vicinity of the sensor more than one peak is expected to be visible on I-V plot. This phenomenon may be used for the simultaneous determination of several toxic gases.

EXPERIMENTAL

Sensor preparations

Nasicon (chemical formula $Na_{2.8}Zr_2Si_{1.8}P_{1.2}O_{12}$) powder was prepared by the conventional solid-state ball milling method[12]. Mixture of chemically pure $NaHCO_3$, ZrO_2, SiO_2 and $NH_4PO_4\cdot3H_2O$ was several times milled and then calcinated at 900°C, and to finalize synthesis fired at 1200°C. Lisicon (chemical formula $Li_{14}Zn(GeO_4)_4$) powders were prepared similarly[13]. Stoichiometric quantities of Li_2CO_3, GeO_2 and ZnO were thoroughly milled and then calcinated for 2 hours at 700°C in platinum boats. The products were reground and fired again for 1 hour at 1100°C to complete the reaction. White, fine-grained powder was obtained in both cases.

Figure 2. Structure of the sensors

The sensor structure made in this way is shown in Fig. 2. Pellets in the form of discs of 12-mm diameter and 1 mm thick were prepared by isostatic pressing and sintering. Electrodes were made by coating opposite pellet faces with the gold paste (ESL 8880) sensor or with the platinum paste (ESL 5542).

Measuring stand

Measurements were conducted in mixtures of high purity gases: nitrogen dioxide, sulfur dioxide, and synthetic air of controlled concentrations. The precision mass flow controllers were used for obtaining gas mixtures composition. A constant gas flow of 100 sccm was maintained. Sensors were placed in a tube furnace in the temperature range from 300°C to 550°C. The measurements were performed using the impedance analyzer SI 1260, the electrochemical interface SI 1287 and a PC computer with suitable software for system control and data acquisition.

Measurements details

Linearly changing voltage excitation signal of symmetrical triangular shape in range from 5 V to –5 V was applied to the sensors terminals (Fig. 3.). The voltage sweep rate was adjusted from 20 to 50 mV/s. Simultaneously the electrical current of the sensor was measured. The sensors response was presented in form of current-voltage plots. Measurements of sensors response in different concentration of NO_2 and SO_2 in balance of the synthetic air were performed. In each case the current-voltage plots of unique shape were obtained. Single peaks were usually visible in the current-voltage plot, however sometimes the plot was more complicated. Impedance measurements were conducted in the frequency range from 100 mHz to 1 MHz with the excitation amplitude of 20 mV.

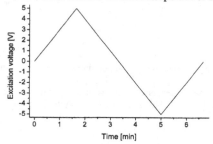

Figure. 3 Triangular shape voltage excitation with sweep rate adjusted to 50 mV/s

RESULTS

Materials investigation

Control of the electrolyte materials preparation process has been provided by X-ray diffraction technique (XRD). This technique allows to determine whether the prepared materials have appropriate crystallographic structure. XRD analysis measurements were carried out using Philips X'Pert diffractometer system. Results show a very good agreement in peak positions and intensity with reference data (Fig. 4).

The electrical conductivity of prepared sensors was determined by means of the ac admittance spectroscopy. In the measured frequency range well-resolved semicircular arc appears in the Nyquist plots. This arc is attributed to the bulk properties of electrolyte. On basis of this plot the resistive component of total impedance was established and used to derive the values of conductivity as the function of temperature. A typical Arrhenius plots for both electrolytes are illustrated in Fig. 5. The conductivity of Nasicon is higher then conductivity reported for Lisicon, what makes Nasicon based sensors very promising for operating at lower temperatures. The activation energies of 0.19 eV for Nasicon and 0.4 eV for Lisicon were resolved.

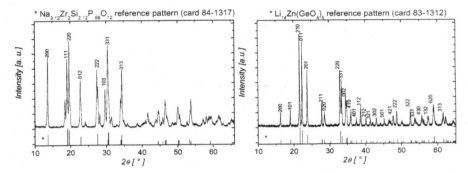

Figure 4. X-ray diffraction pattern of a Nasicon (left) and a Lisicon (right) powders

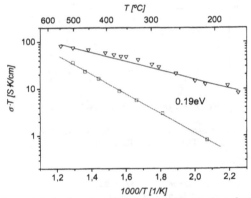

Figure 5. Arrhenius plots of the samples electrical conductivity

Voltammetric sensor response

Performance of Nasicon and Lisicon electrocatalytic sensors exposed to mixtures of nitrogen dioxide of different concentration and synthetic air was investigated at 300°C and 20mV/s sweep rate. For both types of sensors, with the increase of nitrogen dioxide concentration in the atmosphere surrounding the sensor two current peaks are visible. The size of both peaks increases with the concentration of NO_2. In case of Lisicon set of rather plane peaks are observed near ±0.8 V and ±1.9 V (Fig. 6).In case of Nasicon sensor two relatively steep peaks near ±2.3 V are visible (Fig .7).

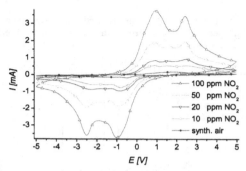

Figure 6. Sensor based on Lisicon response to different NO₂ concentrations (300°C, 20mV/s)

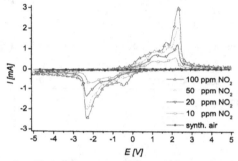

Figure 7. Sensor based on Nasicon response to different NO₂ concentrations (300°C, 20mV/s)

At higher temperatures sensors shows sulfur dioxide sensitivity, while nitrogen dioxide sensitivity is significantly decreased. The current-voltage plots of the sensor exposed to sulfur dioxide at 500°C are presented in Fig. 8.

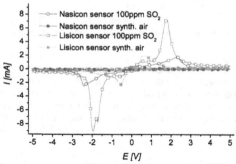

Figure 8. Sensor response to different SO₂ concentrations (500°C, 50mV/s)

Sensor response analysis

Electrocatalytic gas sensor response in the form of current-voltage dependence does not give the possibility to obtain direct information about the type and concentration of gas in the vicinity of the sensors. It is therefore necessary to develop appropriate methods of analysis of the responses of sensors and extract useful information.

A few measures of gas concentration can be used. One simple solution is to analyze the area under the voltammetric curve. This method leads to the calculation of the integral below the surface of the curve in the given limits of integration. The resulting value is expressed in Coulombs, and is connected with the electric charge involved in the electrode reaction. Hence, this value should depend linearly on the concentration of the gas. It is important to find the appropriate limits of integration. Assuming that the operation of the sensor is affected only by one dominant reaction, as the limits of integration would take the whole range of voltage variation (Eg. 3).

$$c_Q = \int_{t(U=0V)}^{t(U=5V)} I dt \; [C] \tag{3}$$

As an example characteristics obtained with this method for sensor based on Lisicon are presented in Fig. 9. In the case of curves obtained in the gas mixture determination of the appropriate limits of integration is difficult or even impossible. Moreover, for tested sensors obtained I-V curves have sometimes complicated shape and there are no easily separated current peaks. Therefore, in practice it is impossible to distinguish parts of the curve associated with the various chemical reactions, occurring for the various components of the mixture of gases. This method is therefore not suitable for the analysis of the sensor responses obtained in the gas mixture.

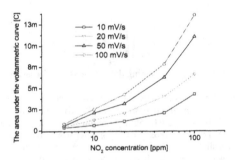

Figure 9. The are under I-V characteristics of sensors based on Lisicon at 300°C

More possibilities gives analysis of a peak current maximum (Eq, 4). Location of the peak can be related to the type of gas, and the size of the peak depends on the value of the measured gas concentrations.

$$c_I = \max I_{peak} \; [A] \tag{4}$$

As an example characteristics obtained with this method for sensor based on Lisicon are presented in Fig. 10.

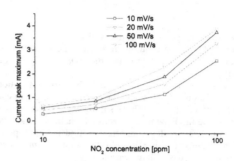

Figure 10. The currant peak maximum of sensors based on Lisicon at 300°C

Studies in a number of gases mixtures have shown that this method is effective only on a limited basis for selected toxic gases. I-V characteristics of sensors often contain a few peaks. It is difficult to determine a way these peaks depend on the concentration of test gas. Location of the peak is not fixed and may depend in general on gas concentration, the temperature and the rate of voltage sweep.

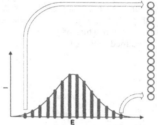

Figure 11 Procedure of obtaining data input vector for advanced processing

For the analysis of the responses of electrocatalytic gas sensors can also be used approach of more advanced data processing methods based on explorative and confirmative techniques as well as artificial neural network. The use of such solutions requires treatment as the output signal of the sensor the whole shape of the I-V response, thus allowing greater use of the information contained therein. In order to verify this approach measurement of a series of sensors responses to various mixtures of toxic gases and synthetic air was carried out. In this way, a set of patterns of individual gas mixtures for various concentrations of NO_2, SO_2 and CO_2 in synthetic air for a range of temperatures was obtained. Obtained curves are divided into 100 points as illustrated in Figure 11. The resulting vector of 100 values of current was then further processed. Positive results of distinguishing the type of gas in the presence of the sensor were obtained with principal components analysis (PCA) and artificial neural network-based adaptive vector quantization algorithm (Learning Vector Quantization - LVQ).

Concept of PCA usage to analyze electrocatalytic gas sensor response is presented in Fig. 12. Plot represented by 100 points is reduced by mathematical computation to significantly smaller number of PC elements. In case of the test data used in all investigated gas composition the 97% of total information (variance) contained in plots can be represented by the 3 first PC elements (Fig. 13).

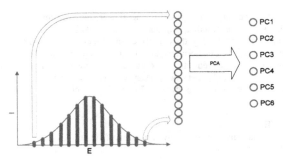

Figure 12 Concept of PCA technique usage

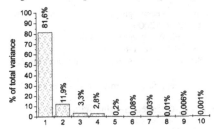

Figure 13 Total variance contained in each PC elements

Those PCs can be used as a new axis in transformed space for a new presentation of measured data. This approach allows showing previously hidden dependencies. This is demonstrated in Fig. 14 (PC1/PC2 representation). Measurements obtained in case of the different gas type are grouped together in easy to separate classes.

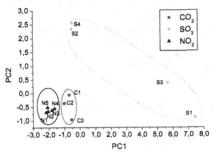

Figure 14. PC1 vs PC2 data presentation

CONCLUSIONS
In this paper the results of cyclic voltammetry applied to gas sensors employing lithium and sodium ion conducting solid electrolytes are presented. It was shown that both investigated sensors could be used for the detection of NO_2 and SO_2. Current-voltage response depends in a unique way on

the type of gas and its concentration exposed to the sensor. Nasicon and Lisicon respond in similar way. Several problems exist in practical applications. The peaks are rather broad and their position depends on the concentration of the gas, thus complicating the deconvolution of different peaks. Proper analysis method should be used. Furthermore, the electrode / electrolyte boundary usually shows high capacitances which lead to undesired capacitive currents during the voltage sweep. In addition, the voltammetric method allows only periodic measurements. Comparatively long times are hence necessary for one complete measuring cycle. This problem might be solved by optimization of excitation voltage shape.

ACKNOWLEDGEMENTS

This work was partially supported by the Polish Ministry of Science and Higher Education under grant No. N N515 243437.

REFERENCES

[1] V. Thangadurai and W. Weppner, Recent progress in solid state oxide and lithium ion conducting electrolytes research, *Ionics*, **12**, 81-92 (2006).

[2] W. F. Chu, V. Thangadurai and W. Weppner, Ionics - a key technology for our energy and environmental needs on the rise, *Ionics*, **12**, 1-6 (2006).

[3] P. Knauth and H. L. Tuller, Solid-state ionics: Roots, status, and future prospects, *J. Am. Ceram. Soc.*, **85**, 1654-80 (2002).

[4] C. O. Park, S. A. Akbar and W. Weppner, Ceramic electrolytes and electrochemical sensors, *J. Mater. Sci.*, **38**, 4639-4660 (2003).

[5] M. J. Madou and S. R. Morrison, Chemical Sensing with Solid State Devices, Academic Press, New York, 1989.

[6] J. Liu and W. Weppner, Θ – sensors: a new concept for advanced solid-state ionic gas sensors, *Appl. Phys. A*, **55**, 250-57 (1992).

[7] M. Liu and L. J. Li, Multi-functional sensor for combustion systems, Patent US5667652(A), 1997

[8] E. L. Shoemaker, M. C. Vogt, F. J. Dudek, and T. Turner, Gas microsensor using cyclic voltammetry with a cermet electrochemical cell. *Sens. Actuators, B*, **42**, 1-9 (1997).

[9] P. Jasinski and A. Nowakowski, Simultaneous detection of sulfur dioxide and nitrogen dioxide by Nasicon sensor with platinium electrodes, *Ionics*, **6**, 230-34 (2000).

[10] G. Jasinski, P. Jasinski, A. Nowakowski, T. Zajt and B. Chachulski, Electrocatalytic nitrogen dioxide sensor, *Proc. SPIE-Int. Soc. Opt. Eng.*, **5505**, 89-94 (2004).

[11] G. Jasinski, P. Jasinski, A. Nowakowski and B. Chachulski, Thick film Lisicon sensor with gold electrodes, *Proc. 28th Int. IMAPS-Poland Conf.*, Wroclaw, 269-72 (2004).

[12] D. H. H.Quon, T. A.Wheat and Nesbitt W., Synthesis, characterization and fabrication of $Na_{1+x}Zr_2Si_xP_{3-x}O_{12}$, *Mater. Res. Bull.*, **15**, 1533-39 (1980),

[13] H. Y.-P. Hong, Crystal structure and ionic conductivity of $Li_{14}Zn(GeO_4)_4$ and other new Li^+ superionic conductors, *Mater. Res. Bull.*, **13**, 117-24 (1978).

WIRELESS CHEMICAL SENSOR FOR COMBUSTION SPECIES AT HIGH TEMPERATURES USING 4H-SIC

Geunsik Lim and Aravinda Kar

Laser-Advanced Materials Processing Laboratory, CREOL, The College of Optics and Photonics, Department of Mechanical, Materials and Aerospace Engineering, University of Central Florida, Orlando, Florida 32816-2700, USA

ABSTRACT

Crystalline silicon carbide (SiC) is an attractive wide bandgap semiconductor material for gas sensor applications in harsh environments because of its high mechanical strength and chemical inertness at elevated temperatures. The optical properties of 4H-SiC can be changed by doping it with appropriate dopant elements to create a dopant energy level that matches with the characteristic emission spectral line of the combustion gas. The radiation emitted by the gas of interest changes the electron density in the semiconductor by the photoexcitation and, thereby, alters the refractive index of the sensor. Since the 4H-SiC substrate inherently acts as a Fabry-pérot interferometer, the experimental data yield an inteferrometric pattern for the reflected power of a He-Ne laser of wavelength 632.8 nm as a function of temperature. The variation of the refractive index has been obtained from this pattern up to 650°C, which provides a mechanism for constructing wireless chemical sensors. A gallium-doped 4H-SiC sensor with dopant energy level $E_V + 0.30$ eV showed a distinct refractive index curve for CO_2, which was different from the curves obtained for NO and NO_2 gases. The dopant energy level is confirmed from optical absorption measurements in the wavelength range of 0.2 to 25 m. The selective changes in the refractive index due to CO_2 indicate that the Ga-doped 4H-SiC substrate can be used as a wireless CO_2 gas sensor.

INTRODUCTION

Emissions control for combustion chambers and turbine systems in power plants require gas sensors that can operate in chemically reactive high temperature environments for both real time monitoring and feedback control of exhaust products. Gaseous species that need to be monitored include hydrogen, hydrocarbons, nitrogen oxides and sulfur oxides. The instrumentation sensors are subjected to the engine's harsh operating environment. These sensors thus offer great promise for monitoring and control of combustion and energy conversion technologies of the future. Moreover, accurate monitoring of the combustor and first-stage turbine components, or their operating environment, is important for preventing component failure and reducing maintenance. This imposes both material and packaging challenges to conventional semiconductor gas sensors[1]. However, some problems remain in the stability of these sensors, and cross sensitivities which are naturally given as the base material have only a general sensitivity to the oxidizing or reducing properties of gases. Silicon-based devices are limited to an operation temperature below 250°C[2] because silicon does not keep its nature as a semiconductor in the region where a lot of electron-hole pairs are thermally generated. One of the failure mechanisms associated with these traditional gas sensors includes, but is not limited to, wirebond failures[2, 3]. Nevertheless, there has been an increasing demand for electronic control concepts occurring from the automotive and power plant industries, where sensor components of such control systems must operate under harsh environment conditions. Sensors of such units have to withstand temperatures up to 600°C or aggressive media like oil, exhaust gases and humidity, as well as mechanical loads like high pressures.

A more convincing and economically more favorable way to cope with higher temperatures and other aggressive environmental conditions is to use new materials. Silicon carbide (SiC) has attracted considerable attention in recent years as a potential material for sensor devices in harsh environments, because it is a wide bandgap and low intricsic carrier concentration semiconductor with better mechanical strength, thermal stability and chemical inertness than many other dielectric materials. The SiC-based sensing system has been shown in various publications as a sensor component in gas sensing applications. SiC exhibits excellent thermal and mechanical properties at high temperature and fairly large piezoresistive coefficients, a combination that makes it well-suited for high temperature electromechanical sensors[4, 5]. Recently, SiC-based resistive device structures have been investigated for their potential use as gas sensing devices at elevated temperatures. Fawcett et al.[6] demonstrated the ability to detect a wide range of H_2 concentration, from 0.33% to 100% H_2 in Ar, a range not seen in catalytic gate Schottky diode or MOSFET sensors at a relatively high temperature of 400°C. Planar NiCr contacts were deposited on a thin 3C-SiC epitaxial film grown on thin Si wafers bonded to polycrystalline SiC substrates. The change of surface conductivity can be measured electronically and correlated to the concentration of the target gases by the 3C-SiC resistive sensor.

In a typical gas detecting components, the sensor material is a catalytic gate field-effect device (metal-insulator-semiconductor structure) with a catalytically active and gas permeable metal such as Pd or Pt. Lolee et al.[7] developed n-type metal-oxide-semiconductor (MOS) capacitive device on SiC (Pt-SiO$_2$-SiC) as the high-temperature gas sensor. The sensor can detect hydrogen under optimum conditions with 1 millisecond at 620°C. Hydrogen molecules dissociate on Pt surface and diffuse through the catalytic metal which result in shifts the capacitance-voltage (C-V) due to the formation of H-induced dipole layers at the Pd-SiO$_2$ interface by hydrogen atoms. Tobias et al.[8, 9] also have demonstrated that the SiC based Schottky diode is able to operate up to 600°C. Spetz et al.[10, 11] developed MISiCFET (Pt/TaSi$_x$/SiO$_2$/6H-SiC) device over a temperature range of 100-700°C. The sensor responds to reducing gases like hydrogen, hydrocarbons and carbon monoxide.

SiC-based solid state gas sensors in aforementioned applications are compromised due to junction breakdown and electromigration[12]. Stable long-term operation is limited to a few days by measurement and temperature permanently induced modifications in the materials. Common gas sensing technology that is usually based on the direct contact of the gas sample with the sensor is very difficult or impossible to use in such cases due to the harsh conditions found in these scenarios. Often the gases to be analyzed are at very high temperatures (up to 1500°C). Furthermore samples may contain extreme dust loads, other reactive, aggressive or condensing species as well as highly sensitive multi-phase mixtures of vapor plus liquid or solid phase particles that shift their phase partitioning on any temperature/pressure/composition change. These boundary conditions make it very difficult, if not impossible, to condition (i.e. sample, filter or cool) the gases for the measurement and the transport them to the sensor.

Researchers have turned to optical detection mechanism for providing a robust chemical sensing solution in the hazardous and severe environments. Especially, laser diagnostics are facilitating improved understanding of a wide variety of combustion phenomena which, in turn, will lead to improve efficiency and cleanliness in the energy conversion devices. Compared to conventional electronic sensors, optical sensors have many advantages including small size, light weight, high sensitivity, high reliability, corrosion and oxidation resistance. Optical sensors can also survive at much higher temperatures and pressures than conventional high temperature and pressure sensors. Optical sensors to detect combustion gases, for instance CO, CO_2, and NO$_x$, have been designed for use up to 650°C in harsh environments. Temperature, pressure and chemical sensor using single crystalline SiC substrates for harsh environments have been developed with optically reflective and interferometric techniques[13-15]. Engineers have observed the inherent interferometric capability of SiC and analyzed its applicability as a wireless optical sensor to detect combustion species at high temperatures and high pressures. Prior works include using thin films of SiC to act as Fabry-Pérot etalons to form high pressures (up to 600 psi) and high-temperature (up to 500°C) sensors[13-15]. In particular, high-temperature gradients and fast temperature/pressure temporal effects can cause stress fields at the SiC thin film-gas interface causing deterioration of optical properties (e.g., interface reflectivity) required to form a quality Fabry-Pérot etalon need for sensing based on SiC film refractive index change. Such applications may range from in situ monitoring of combustion processes (helping to make such processes more efficient and controlling unwanted emissions) to scientific applications (such as un-cooled radiation detectors in new generations of space-craft).

In this work, we utilized a doped 4H-SiC gas chemical sensor to successfully demonstrate its survivability in the high temperature environment. A continuous wave He-Ne laser of wavelength 632.8 nm is used to understand the changes in the refractive index of a composite layer at the sensor-gas interface in the presence of different combustion gases. The SiC sensor is inherently a Fabry-Perot interferometer enabling optical measurements pertaining to interference patterns that were analyzed to calculate refractive index. The detection of changes in the refractive index by optical techniques is a new approach to measure the gas concentration with improved accuracy which can be used as an optical device for remote gas sensor. Such sensors were fabricated by doping a single crystal n-type 4H-SiC substrate with gallium dopant element and were tested for carbon dioxide (CO_2) detection at high temperatures up to 650°C.

EXPERIMENTAL PROCEDURE
SENSOR FABRICATION

As for most combustion-product sources such as H_2O, CO_2, CO and NO_x, the infrared emission is consequent to changes in the energy contained in the molecular vibrations and rotations. In the infrared region, the principal or fundamental bands for these molecules arise from vibrational transitions; the very strong band at 4.21 m is due to the characteristic vibrational mode of CO_2[16]. For combustibles, sensitive and selective detection of the gas is possible in the spectral range between 4 and 5 m due to the emission line of CO_2 is present in the absence of strong water absorption which is present nearly everywhere in the infrared spectral region. Therefore, an optical gas sensor is developed by doping a 4H-SiC with an appropriate dopant to match the dopant energy level with the energy of the radiation emitted by the combustion gas of interest. The wavelength from characteristic emission spectra correspond to the photon energy $E = hc/$, where h is Planck's constant, c is the speed of light in vacuum. Lebedev et al.[17, 18] reported a set of dopant elements and their energy levels in SiC, which showed that gallium atom produces acceptor energy levels of 0.30 eV in 4H-SiC. Therefore the Ga element will be incorporated into 4H-SiC to fabricate sensors for the CO_2 gas.

Gallium was doped in n-type 4H-SiC substrate of length, width and thickness 1, 1 and 0.0375 cm respectively with laser doping system consisting of a vacuum chamber and a bubbler as shown in Fig. 1. The sample was cleaned by soaking in H_2O_2:H_2SO_4 (1:1 by volume) solution for 15 minutes, de-ionized water rinsing and buffered oxide etch (BOE) dipping. The cleaned sample was placed in a doping chamber and the chamber was pumped down to 1 mTorr vacuum. Triethylgallium, $(C_2H_5)_3Ga$, was used as a Ga dopant precursor, which was heated in a bubbler immersed in a waterbath maintained at 100°C. The precursor vapor was delivered to the doping chamber by Ar carrier gas, while the SiC substrate was heated with a continuous wave (CW) Nd:YAG (= 1.064 m) laser beam. The laser processing parameters for the laser doping of SiC with Ga and Al are the same (laser power: 10.5 W, focal length: 150 mm, spot size: 200 m, scanning speed of the micro-stage: 0.8 mm/sec). After the laser-doping experiments, the sample was cleaned with a 45 wt.% KOH solution and rinsed with acetone, methanol and DI water.

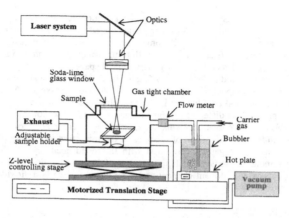

Fig. 1. Experimental setup for laser doping of 4H-SiC with Ga to fabricate gas sensor.

SENSOR TESTING

A CW He-Ne laser (632.8 nm) beam of output power up to 15 mW was used to measure the optical response of the Ga-doped 4H-SiC sensor. The laser beam reflected off the sensor was directed to a photodetector as shown in Fig. 2. Computer data acquisition software was used to record the temperature of the stainless pedestal which contains CO_2 gas inside and the power of the reflected He-Ne beam. One of the important optical properties of doped SiC is that it inherently acts as a Fabry-Pérot interferometer in which the laser beam reflects off both the top and bottom inner surfaces of the sensor. Such reflections create a multibeam interference pattern, which is analyzed to calculate the change in the refractive index of the sensor by a particular combustion gas. Experiments were conducted with the as-received and Ga-doped 4H-SiC respectively to analyze and compare the effect of doping of the sensor.

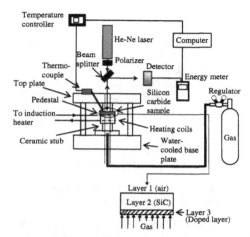

Fig. 2. Experimental setup to obtain the interferometric pattern of the sensor for different combustion species.

RESULTS AND DISCUSSION

Wireless optical sensors have been fabricated by laser doping of a 4H-SiC substrate with gallium which is capable of detecting CO_2 gas. The sensor operates on the semiconductor optics principle; there is an energy gap in the 4H-SiC semiconductor created by the dopant atoms which directly relates to an emission spectral line of the gas interest. The electrons in the doped sample, excited by the photons emitted by the gas, consequently change the density of the electrons in various energy states, which in turn affects the refractive index and then the reflectance of the sensor. The optical response of the sensor for the reflected power of a probe beam, particularly with the He-Ne laser in this study, is an interference pattern which consists of multiple thin-film layers each with a different refractive index. A light beam incident perpendicular to the surface of the multilayer stack is used to propagate information through the structure. The reflectance values (or alternatively, reflection coefficients) of the light incident can be used as the general measurement of the outputs. Since the reflected beams different from each other in both amplitude and phase with a reflected beam and a refracted beam that enters the sensor material, the overall reflection of the beam off the sensor region will be determined by the multiple-beam interference of the infinite series of reflected component waves as shown in Fig. 3. Analyzed from these complementary Airy patterns[13-15], the doped 4H-SiC sample that has the refractive indices can be used to identify the gas.

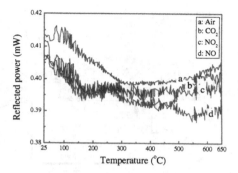

Fig. 3. Reflected power of a He-Ne laser beam from an as-received 4H-SiC substrate for different gases; air, carbon dioxide, nitrogen dioxide and nitrogen monoxide from 25 to 650°C.

While Fig. 3 shows the optical response of an as-received SiC substrate, which establishes the base line (initial) properties of the sample, the interference patterns of a Ga-doped 4H-SiC sample are shown in Fig. 4. This Ga-doped sample acts as a sensor for the CO_2 gas. Gallium produces an acceptor energy level in 4H-SiC corresponding to the CO_2 emission spectral line 4.21 m (i. e., 0.30 eV).

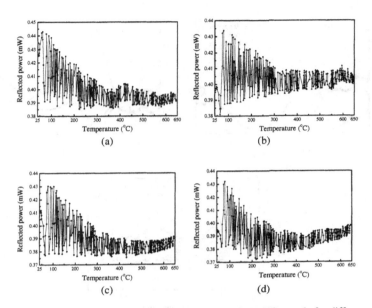

Fig. 4. Reflected power of a He-Ne laser beam from a Ga-doped 4H-SiC sample for different gases; (a) air, (b) carbon dioxide, (c) nitrogen monoxide and (d) nitrogen dioxide from 25 to 650°C.

The dopant energy level was verified by optical absorptance measurements in the wavelength range 2.5 to 25 m using a Bruker Vertex 70 FTIR spectrometer equipped with a Helios infrared micro-sampling accessory. The Helios combines a rapid scan speed of 100 spectra per second with a simple infrared microscope (resolution 250 μm). The absorptance of the Ga-doped sample (0.32) is higher than that of the as-received sample (0.12) at 4.21 m as shown in Fig. 5(c), and this wavelength corresponds to the Ga dopant energy level (0.30 eV) in 4H-SiC. However, the Ga-doped sample exhibits another absorption peak at the wavelength 4.63 m. The absorption coefficients of the as-received sample are 0.360 mm^{-1} at 4.21 m and 0.358 mm^{-1} 4.63 m. On the other hand, the absorption coefficients of the doped sample are found to be 1.601 mm^{-1} at 4.21 m and 1.461 mm^{-1} 4.63 m, which indicate that the Ga dopant has increased the absorption characteristics of the substrate.

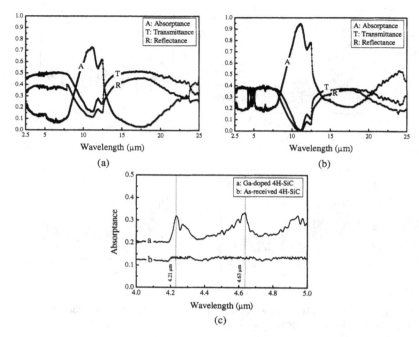

Fig. 5. Optical properties of (a) as-received, (b) Ga-doped 4H-SiC in the wavelength range 2.5 – 25 m and (c) comparison of absorptance within 4-5 m.

To verify the selectivity of the CO_2 gas, the Ga-doped 4H-SiC sensor was demonstrated for combustion gases such as CO_2, NO and NO_2. The refractive index of the gas-excited doped 4H-SiC (n_g) depends on the wavelength ( ), temperature (T), gas concentration (c) and gas type, i.e., $n_g = n_g(,T,c)$. The deviation of n_g from the refractive index of the doped sample (n_a) in air, i.e., $n_a = n_a(,T)$, which is expressed as $n = n_g(,T,c) - n_a(,T)$, is an important parameter that shows the selectivity of the sensor for a particular gas.

The CO_2 gas has a markedly different curve from NO and NO_2 gases, as is shown in the refractive index ($n_{Ga\text{-}doped} = 3.530$ at 650°C) and the change in refractive index (n) of the Ga-doped sample as shown in Figs. 6(a) and (b). In this study of development sensors the electronic transition brought about by the photoexcitation of the CO_2 gas only occurs from the lower to upper energy level, as the Ga dopant energy level in 4H-SiC ($E_V + 0.30$ eV) matches with the photon energy of the CO_2 gas (0.30 eV corresponding to 4.21 m wavelength). Fig. 6(b) shows that the value of n is unique for the CO_2 gas. Therefore, the Ga-doped 6H-SiC can be used as a CO_2 gas sensor. The dopant energy level in SiC of Ga is verified by the absorption spectra which were obtained at room temperature, shows that the absorption peaks exactly match with the photon energy of the CO_2 gas. Furthermore, at high temperatures, the gas might have an emission spectral line matching with the Ga dopant energy level in 4H-SiC. So we find that Ga-doped 4H-SiC can be instrumental in the detection of CO_2 gas, since the refractive index data thus shows that the CO_2 gas affects the index of the doped sample selectively in conjunction with those high temperatures.

In the spectral range of 0.78 – 3.0 µm, IR detectors are normally used for applications as diverse as fiber optic communications, agricultural sorting, environmental monitoring, and chemical analysis. Further into the IR region (from 2 – 5 µm), applications for IR detectors include non-contact temperature sensing, thermal imaging, and gas analysis for pollution control. The 3 – 5 µm band is more appropriate for hotter objects, or if sensitivity is less important than contrast. It has advantages of lower ambient and background noise. This spectral region is thus optimal for such applications as thermal imaging, non-contact temperature sensing, security sensing, and environmental monitoring[19]. The Ga-doped 4H-SiC sensor can provide best performance at these wavelengths.

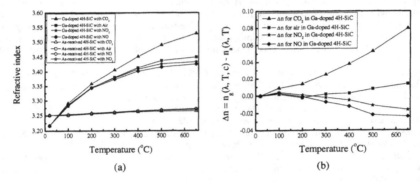

(a)   (b)

Fig. 6. (a) Refractive indices and (b) changes in refractive indices of Ga-doped 4H-SiC sensors for different combustion gases.

CONCLUSIONS

The effects of gas on the optical properties of 4H-SiC sensor have been studied as a function of temperatures and pressures. The reflected power of the sensor exhibited unique interference patterns for different gases. These patterns have been utilized to determine the refractive index of a composite layer at the sensor-gas interface in order to develop a SiC sensor to identify the combustion gases. The radiation from the hot gases changes the electron or hole density in the doped semiconductor by the photoexcitation of electrons from the valence band to the dopant energy level, if the SiC is doped with Ga dopant material to create energy gaps corresponding to the photon energy 0.30 eV for CO_2. The CO_2 exhibits distinct refractive indices for different partial pressures of CO_2 in the mixture compared to the refractive index of the gas, which indicates that the refractive index can be used to sense CO_2 gas.

REFERENCES

[1] J. Casady and R. Johnson, Status of Silicon Carbide (SiC) as a Wide-Bandgap Semiconductor for High-Temperature Applications: A Review, *Solid-State Elecctron.*, **39**, 1409-1422 (1996).

[2] L. Chen, G. Hunter, P. Neudeck, and D. Knight, Surface and Interface Properties of PdCr/SiC Schottky Diode Gas Sensor Annealed at 425°C. *Solid-State Electron.*, **42**, 2209-2214 (1996).

[3] A. Spetz, P. Tobias, A. Baranzahi, P. Martensson, and I. Lundstrom, Current Status of Silicon Carbide Based High-Temperature Gas Sensors. *IEEE Trans. Electron Devices*, **46**, 561-566 (1999).

[4] G. Hunter, L. Chen, P. Neudeck, D. Knight, C. Liu, Q. Wu, and H. Zhou, Development of High Temperature Gas Sensor Technology, *42nd Turbo-Expo 97*, Orlando, USA, June 1997.

[5]R. Okojie, Single-Crystal Silicon Carbide MEMS: Fabrication, Characterization, and Reliability, Chapter 7, MEMS: Design and Fabrication, edited by G. Mohamed, Taylor and Francis, New York, 2005, pp. 7.1-7.6.

[6]T. Fawcett, J. Wolan, A. Spetz, R. Myers, and S. Saddow, Thermal Detection Mechanism of SiC Based Hydrogen Resistive Gas Sensors, *Appl. Phys. Lett.*, **89**, 182102 (2006).

[7]R. Loloee, B. Chorpening, S. Beer, and R. Ghosh, Hydrogen Monitoring for Power Plant Applications Using SiC Sensors, *Sens. Actuat. B*, **129**, 200-210 (2008).

[8]P. Tobias, B. Golding, and R. Ghosh, Interface States in High-Temperature Gas Sensors Based on Silicon Carbide, *IEEE Sens. J.* **3**, 543 (2003).

[9]Tobias P, Hui H, Koochesfahani M and Ghosh R N, *Proc. IEEE Sensors* (Vienna, Austria, Oct. 2004)

[10]A. Spetz, P. Tobias, L. Uneus, H. Svenningstorp, L. Ekedahl, and I. Lundstrom, High Temperature Catalytic Metal Field Effect Transistors for Industrial Applications, *Sens. Actuators B*, **70** 67-76 (2000).

[11]A. Spetz, L. Uneus, H. Svennningstorp, P. Tobias, L. Ekedahl, O. Larsson, A. Goras, S. Savage, C. Harris, P. Martensson, R. Wigren, P. Salomonsson, B. Haggendahl, P. Ljung, M. Mattsson, and Lundstrom I, SiC Based Field Effect Gas Sensors for Industrial Applications, *Phys. Stat. Sol. A*, **185** 15-25 (2001).

[12]N. Wright and A. Horsfall, SiC Sensors: A Review, *J. Phys. D: Appl. Phys.*, **40**, 6345-6354 (2007).

[13]S. Dakshinamurthy, N. Quick, and A. Kar, SiC-based Optical Interferometry at High Pressure and Temperature for Pressure and Chemical Sensing, *J. Appl. Phys.*, **99**, 094902 (2006).

[14]A. Chakravarty, N. Quick, and A. Kar, Decoupling of Silicon Carbide Optical Sensor Response for Temperature and Pressure Measurements, *J. Appl. Phys.*, **102**, 073111 (2007).

[15]S. Dakshinamurthy, N. Quick, and A. Kar, Temperature-dependent Optical Properties of Silicon Carbide for Wireless Temperature Sensors, *J. Phys. D: Appl. Phys.*, **40**, 353-360 (2007).

[16]U. Willer, M. Saraji, A. Khorsandi, P. Geiser, and W. Schade, Near- and Mid-Infrared Laser Monitoring of Industrial Processes, Environment and Security Applications", *Optics and Lasers Eng.*, **44**, 699-710 (2006).

[17]A. Lebedev, Deep-Level Defects in SiC Materials and Devices, Chapter 4, Silicon Carbide: Materials, Processing, and Devices, edited by Z. Feng and J. Zhao, Vol. 20 in series on Optoelectronic Properties of Semiconductors and Superlattices, Taylor and Francis, New York, 2004, pp. 121-163.

[18]A. Lebedev, Deep Level Centers in Silicon Carbide: A Review, *Semiconductors*, **33**, 107-130 (1999).

[19]J. Lloyd, Thermal Imaging Systems, Plenum Press, New York, 1975, pp. 166-184.

HIGH TEMPERATURE ACOUSTIC WAVE GAS SENSOR USING LANGASITE CRYSTAL RESONATOR

Hongbin Cheng, Lifeng Qin, Qing-Ming Wang[*]
Department of Mechanical Engineering & Materials Science, University of Pittsburgh, Pittsburgh, PA 15261, USA

ABSTRACT:
    Inspired by recent development of one-dimensional nanoscale materials with large specific surface area and intriguing properties, in this study c-axis vertically aligned ZnO nanorod arrays were synthesized on the langasite thickness shear mode bulk acoustic wave resonator through a simple hydrothermal route for potential gas sensing application. The ZnO nanorods are characterized by X-ray diffraction (XRD) and scanning electron microscopy (SEM) for phase and microstructure identification. The nanorods have a diameter of 30 to 100 nm and a length of about several hundred nanometers. The langasite high temperature gas sensor fabricated from ZnO nanorod arrays showed a high sensitivity to NO_2 and NH_3 at 300°C. The results demonstrate that the use of the ZnO nanorod arrays on langasite acoustic wave resonator can greatly enhance the sensitivity and sensor response speed due to the fast surface/interface reaction and improved surface characteristics.

INTRODUCTION

    Zinc oxide is one of the earliest and most widely used metal-oxide semiconducting gas-sensing materials due to its high mobility of conduction electrons, low cost, and good flexibility in fabrication [1-3]. However, the sensing performance is often limited by the grain sizes, surface states, and gas adsorption and dissociation rate and the diffusion rate in the thin film materials [4]. The emergence of nanoscale science and technology in recent years is making a significant impact on gas sensors. Novel nanoscale sensitive structures show great promise as they have faster response and higher sensitivity than conventional planar sensor configurations, due to their smaller dimensions combined with dramatically increased sensing surface and strong binding properties [5,6]. To date, various one-dimensional ZnO nanostructures have been synthesized in forms of nanowires, nanobelts, micro- and nano-tubes, and nanocones [7-9]. Those ZnO nanostructures have a high surface to volume ratio, and have been regarded as a promising material to improve the gas sensing performance. Recently, gas sensors based on ZnO nanostructures have been demonstrated for toxic and combustible gas sensing, and these sensors show increased sensitivity and accuracy [10].

    Acoustic wave gas sensors using metal oxide thin film as the sensitive interface have been studied for detection of specific gases at low level concentration in the past years due to high sensitivity and structural simplicity [11]. Thickness shear mode (TSM) quartz crystal resonator is the most commonly used device for acoustic wave gas sensors, which detect the gas concentration through the measurement of resonance frequency shift induced by the mass change of active layer due to gas adsorption and reaction. However, for high temperature (>350°C), quartz device is limited due to high electrical loss and α-β phase transformation at 573°C [12].

---

[*] Corresponding author, qmwang@engr.pitt.edu

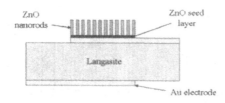

Figure 1. Schematic of langasite gas sensor coated with ZnO nanorods arrays

Figure 2. Characterization of ZnO nanorod arrays (a) XRD, (b) & (c) SEM images: top and cross section

Recently, langasite ($La_3Ga_5SiO_{14}$, LGS) crystal has become a promising piezoelectric material for acoustic devices at high temperature due to its superior characteristics compared with quartz: high bulk resistance (low electrical losses) at high temperature, up to about six times higher electromechanical coupling coefficient than quartz, about 25% reduction in phase velocities with respect to quartz, and existence of temperature compensation orientations. Langasite does not undergo phase transformations up to the melting temperature of 1470°C [13]. Measurements of its resonant frequencies have been performed at temperatures as high as 1000°C. This could allow its use in monitoring high temperature processes where quartz would be not applicable.

Therefore, in this study, we present our recent study on the fabrication and characterization of the ZnO nanorod gas sensor using thickness shear mode (TSM) langasite resonators at high temperature conditions. *c*-axis vertically aligned ZnO nanorod arrays were synthesized on the langasite resonator by using an ultrathin ZnO seed layer through a simple hydrothermal route. Those nanowires are characterized by X-ray diffraction (XRD) and scanning electron microscopy (SEM). NO_2 and NH_3 were particularly used to investigate the sensing performance of langasite BAW gas sensors with ZnO nanorods sensitive layer at 300°C. The results indicated the use of the ZnO nanorod arrays on acoustic wave resonator can greatly enhance the sensitivity due to the fast surface/interface reaction and surface roughness.

EXPERIMENTAL PROCEDURE

The thickness shear mode (TSM) langasite gas sensor with ZnO nanorods sensitive layer is schematically shown in Figure 1. The sensor started from 6 MHz Y-X cut polished langasite crystal (Fomos-Materials Company, Russia) with gold electrodes. ZnO nanorods were grown on one side of the

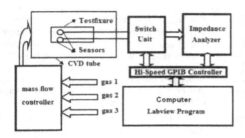

Figure 3. Schematic of test system set-up

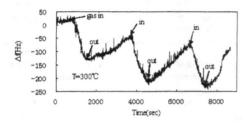

Figure 4. Time cycling response to NO₂ at 300°C

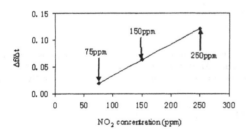

Figure 5. Frequency sensitivity S_f vs. NO₂ concentration at 300°C

resonator according to a two-step hydrothermal method[14]. The morphology and structure were characterized using scanning electron microscopy (SEM) and x-ray diffraction (XRD).The XRD pattern in Figure 2(a) shows that the grown ZnO nanorod arrays can be indexed to the wurtzite phase of ZnO with lattice constants of a=0.325 nm and c=0.520 nm. Only one sharp peak ZnO (002) is found at 34.32 degree position with strong relative intensity. The absence of peaks for other ZnO orientation indicates that ZnO nanorods have a preferred orientation along the c-axis. Figure 2(b) and (c) show typical scanning electron microscopy (SEM) images of ZnO nanorod arrays. The nanorods are found to be uniformly distributed and well aligned normal to the surface of the substrate. In addition, the nanorods have a diameter ranging between 50 and 100 nm, and a length of about 1.5 um.

Figure 3 shows the test system set-up. The LGS resonator mounted on the ceramic plate was placed in the quartz tube of 2 inch diameter and 12 inch length in a CVD system (Tek-Vac Industries, Inc). The impedance spectrum of resonators was achieved through an impedance analyzer (4294A, Agilent), which was used to extract the fundamental resonance frequency through data fitting method. A computer with Hi-Speed GPIB Controller (GPIB-USB-HS, NI) and program (Labview7.0, NI) were used for automatic measurement and data saving. The sensor was investigated in NO₂ and NH₃ with various concentrations balanced in dry air with a total flow rate of 200 sccm at 300°C. Here 300°C was selected due to the frequency of LGS resonator that we bought was most stable at this point. Three mass flow controllers were used to limit the gas flow to a certain rate.

RESULTS AND DISCUSSION

Figure 4 shows the time-cycling response of the gas sensor to 250ppm NO₂ in dry air at the optimal operating temperature 300°C. First, 200 sccm dry air was dlievered through the chamber until the langasite resonator's frequency change become smaller than 1 Hz/min. In this state, we can say that the langasite resonator reaches a balance. Usually, this process takes about 3 hours. Then, 250ppm NO₂ in dry air was applied to the test system for 10 mins with the same flow rate as the previous step. Thereafter, 200

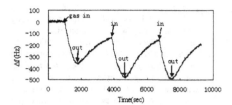

Figure 6. Time cycling response to NH₃ at 300℃

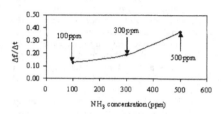

Fig.7. Frequency sensitivity S_f vs. NH₃ concentration at 300℃

sccm dry air was used again to recover the device. All the processes were undertaken at 300℃ and controlled humidity environment. The frequency of the ZnO nanorod arrays sensors reduces with the presence of NO₂. The reversible cycles of the response curve indicate a stable and repeatable operation of gas sensing. However, it is found that the desorption process of NO₂ is much slower than adsorption process.

A frequency-changing rate $S_f = |\Delta f / \Delta t|$ is defined to reflect the sensor sensitivity [15]. The larger S_f value means the bigger frequency shift per unit time and thus the higher sensitivity. Figure 5 shows the corresponding relationship between S_f and the NO₂ gas concentration. It is found that the frequency-changing rate increases with increasing NO₂ concentration.

The response of ZnO gas sensor to NH₃ gas was also tested. Figure 6 and 7 shows the response to time cycle and different concentration at 300℃. The sensitivity S_f also increases with increasing NH₃ concentration.

We can understand the sensing mechanism of ZnO nanorods layer based on the following considerations. In the NO₂ detection, NO₂ gas molecules can be adsorbed or can interact with the oxygen adsorbed onto the surface of the ZnO nanorods layer according to the following reactions [16].

$$NO_{2g} + e^- \rightarrow NO_2^- \tag{1}$$
$$NO_{2g} + O_2^- + 2e^- \rightarrow NO_2^- + 2O^- \tag{2}$$

The adsorbed ions NO_2^- are desorbed as NO₂ gas when NO₂ flow is stopped and consequently in this process a recovery of the initial conditions takes place. Moreover, those oxygen vacancies on the surface of ZnO nanorods might also involve in the processes of physical or/and chemical adsorptions of NO₂ [17, 18]. These oxygen vacancies are the chemisorption sites:

$$NO_2(g) + V_O \rightarrow (V_O - O_{ads}) + NO(g). \tag{3}$$

Oxygen desorption occurs by the reaction,

$$2(V_O - O_{ads}) \rightarrow 2(V_O)_s + O_2(g). \tag{4}$$

The existence of new NO_2^- and NO species on the ZnO surface indicates the reaction between NO₂ and ZnO, which leads to the mass change of the surface layer of the acoustic wave resonator. Furthermore, the possible ZnO/langasite interfacial stress changes may also affect in the resonant frequency of the device. However, the precise verification of this interfacial effect is difficult due to the complex surface states of ZnO-langasite sensor. Therefore, the high sensitivity of acoustic wave gas sensors using ZnO nanorods as the active layer is achieved due to the large surface to volume ratio associated with the nanostructured ZnO layer. At this scale, the vertically aligned morphology of ZnO nanorods provides not only a variety of channels for surface reactions between chemisorbed oxygen species and target gases, which make more surface exposed to the testing gas, but also more oxygen vacancies compared with the flat thin film. As a result, the performance of acoustic wave gas sensor using nanostructured ZnO layer was improved as compared with the use of flat thin film layer.

As the sensing mechanism for NH_3 gas, it also is a process of physical and chemical adsorption of NH_3 gas molecules on the ZnO nanostructures. Besides NH_3 molecules are adsorbed on the sensitive layer at a short time due to the large surface area of ZnO nanorods to cause mass loading change, chemical reactions at the surface of ZnO nanorods between the reducing gas NH_3 and the negatively charged O^- ions may also contribute to the frequency response of the acoustic wave sensors, as was described by [19]:

$$2NH_3 + 3O^- \rightarrow 3H_2O + N_2 + 3e^- \qquad (5)$$

The existence of new H_2O and N_2 species on the ZnO surface indicates the reaction between NH_3 and ZnO, which also leads to the mass change of the surface layer of the acoustic wave resonator.

CONCLUSIONS

In summary, ordered ZnO nanorod arrays were grown on the langasite thickness shear mode resonator by two-step hydrothermal method at low temperature. The acoustic wave gas sensor using ZnO nanorod arrays as the sensitive interface exhibited good gas sensing properties to NO_2 and NH_3. The reproducibility is good since the metal oxides are very stable at the test temperature. The response time of the sensor is relative fast due to the large surface area of ZnO nanorods and reversible surface physical/chemical reactions. In addition, this work demonstrated that the combination of nanorod arrays with langasite thickness shear mode acoustic wave resonators provides a promising high temperature gas sensing platform with both high sensitivity and enhanced response speed. Further sensor text at wide temperature are undergoing and will be reported later.

ACKNOWLEDGEMENTS

The authors would like to acknowledge the financial support from National Science Foundation (NSF) under Grant No.CMMI-0510530.

REFERENCES
[1] T. Seiyama, A. Kato, A. Kato and M. Nagatani, A New Detector for Gaseous Components Using Semiconductive Thin Films, Anal. Chem. 34 (1962), pp. 1502–1503.
[2] M. Aslam, V.A. Chaudhary, I.S. Mulla, S.R. Sainkar, A.B. Mandale, A.A. Belhekar, K. Vijayamohanan, A highly selective ammonia gas sensor using surface-ruthenated zinc oxide, Sens. Actuators 75 (1999) 162–167.
[3] B. Rao, Zinc oxide ceramic semiconductor gas sensor for ethanol vapor, Mater. Chem. Phys. 64 (2000) 62–65.
[4] G. S. T. Rao, and D. T. Rao, Gas sensitivity of ZnO based thick film sensor to NH_3 at room temperature, Sens. Actuators B 55, pp. 166-169, 1999.
[5] Z. Zhang; H. Chen; J. Zhong; Y. Chen; Y. Lu; ZnO Nanotip-based QCM Biosensors" IEEE International Frequency Control Symposium and Exposition, pp. 545 – 549 (2006)
[6] C.-Y. Lu; S.-P. Chang; S.-J. Chang; T.-J. Hsueh; C.-L. Hsu; Y.-Z. Chiou; I-C. Chen; ZnO Nanowire-Based Oxygen Gas Sensor, IEEE Sensors Journal Vol. 9, Issue 4, pp.485 - 489 (2009)
[7] M.H. Huang, Y.Y. Wu, H.N. Feick, N. Tran, E. Weber and P.D. Yang, Catalytic Growth of Zinc Oxide Nanowires by Vapor Transport, Adv. Mater. 13, p. 113(2001).
[8] Zhong Lin Wang, Nanostructures of Zinc oxide, Materialstoday, 26-33 (2004)
[9] H. Cheng, J. Cheng, Y. Zhang, and Q.-M. Wang, Large-scale Fabrication of ZnO Micro-and Nano-Structures by Microwave Thermal Evaporation-Deposition, Journal of Crystal Growth, Vol. 299, Issue 1, pp. 34-40, 2007

[10] Z.P. Sun, L. Liu, L. Zhang, and D. Z. Jia, Rapid sythesis of ZnO nano-rods by one-step, room temperature, solid-state reaction and their gas-sensing properties, Nanotechnology 17, pp. 2266-2270, 2006.

[11] C.-Y. Shen, S.-Y. Liou, Surface acoustic wave gas monitor for ppm ammonia detection, Sensors and Actuators B: Chemical, Vol. 131, Issue 2, pp. 673-679 (2008)

[12] J. Hornsteiner, E. Born, G. Fischerauer, E. Riha, Surface acoustic wave sensors for high temperature applications, IEEE Int. Freq. Countr. Symp., 1998, pp.615–620.

[13] J.A. Thiele, M. Pereira da Cunha, High temperature LGS SAW gas sensor, Sensors and Actuators B: Chemical, Vol.113 (2006) 816-822.

[14] Lori E. Greene, Matt Law, Dawud H. Tan, Max Montano, Josh Goldberger, Gabor Somorjai, and Peidong Yang, General route to vertical ZnO nanowire arrays using textured ZnO seeds, NANO letters 5, pp. 1231-1236, 2005.

[15] J. Hu, F. Zhu, J. Zhang, H. Gong, A room temperature indium tin oxide/quartz crystal microbalance gas sensor for nitric oxide, Sens. and Actuators B 93, pp.175-180, 2003

[16] B. Ruhland, Th. Becker, G. Muller, Gas-kinetic interactions of nitrous oxides with $SnO2$ surfaces, Sens. and Actuators B 50, pp. 85–94, 1998.

[17] C. S. Rout, K.. Ganesh, A. Govindaraj, C.N.R. Rao, Sensors for the nitrogen oxides, NO_2, NO andN_2O, based on In_2O_3 and WO_3 nanowires, Appl. Phys. A 85, 241–246 (2006).

[18] L. Francioso, A. Forleo, S. Capone, M. Epifani, A.M. Taurino, P. Siciliano, Nanostructured In_2O_3-SnO_2 sol-gel thin film as material for NO_2 detection, Sens. Actuators B 114, 646 (2006)

[19] M.S. Wagh, G.H. Jain, D.R. Patil, S.A. Patil, L.A. Patil, Modified zinc oxide thick film resistors as NH_3 gas sensor, Sens. and Actuators B 115, pp.128–133, 2006.

SYNTHESIS OF (La,Nd):Y$_2$O$_3$ AND (La,Yb):Y$_2$O$_3$ LASER CERAMICS AND THEIR OPTICAL PROPERTIES

Yihua Huang Dongliang Jiang
The State Key Lab of High Performance Ceramics and Superfine Microstructure, Shanghai Institute of
Ceramics, Chinese Academy of Sciences, Shanghai 200050, China
Email: dljiang@sunm.shcnc.ac.cn

ABSTRACT:

Nd^{3+}, La^{3+}:Y$_2$O$_3$ and Yb^{3+}, La^{3+}:Y$_2$O$_3$ transparent ceramics were reported. La^{3+} is used as a sintering additive, while Nd^{3+} and Yb^{3+} are doped as laser activators. Initially, yttria and appropriate amount of lanthanum oxide powder was dissolved in nitric acid to form a mixture of nitrate solutions. Then ammonia solution was added dropwise under stirring. After aging, the precursor gel was washed by deionized water for several times to remove the byproducts. After freeze-drying and calcinations, the powder was uniaxially compacted into pellets, followed by sintering in vacuum. The microstructure and the optical properties of the obtained transparent yttria ceramics were investigated. Fluorescence analysis shown that the obtained materials are suitable for laser applications.

1. INTRODUCTION

Neodymium and Ytterbium ion are the most hopeful ions that can be used in laser. Nd ion is well accepted in laser application(1, 2). While Yb is also a very attractive dopant for efficient diode pumped solid-state lasers due to its very simple energy level(3): the ^{2}F$_{7/2}$ ground state and the ^{2}F$_{5/2}$ exited state. There is no excited state absorption which might reduce the effective laser cross section, no up conversion, no concentration quenching. Thus, Yb^{3+} shows high quantum efficiency, weak non-radiative transitions.

The fabrication of highly transparent ceramics for laser application(4, 5), which is better than single crystal, has been realized. Many cubic garnet like YAG, LuAG and sesquioxide transparent polycrystalline ceramics have shown efficient laser emission properties and got laser output(6, 7).
There are several ways to fabricate fully densified yttria transparent ceramics, such as sintering with additives. Some additives can form liquid phase during the sintering of yttria. Lefever (8) et al fabricated the first transparent yttria ceramic in 1967. They used hot pressing at 950□ under 70MPa using LiF as sintering additive. The addition of LiF was responsible for the formation of liquid phase during sintering. SrO(9) was found to be a good candidate for accelerating densification process. The achievement of transparency was attributed to the creation of point defects, especially oxygen vacancies, which helped to accelerate pore removal during the sintering process. In 2002, Ikegami (10) et al fabricated transparent yttria ceramics from low temperature synthesized yttrium hydroxide precursor. Sulfate ion was used as an additive, and liquid phase sintering may occur because considerable pores were eliminated in the final sintering.

Some additives can inhibit the grain growth in the sintering of transparent yttria ceramics by solute drag mechanism (11). Greskovich (12) fabricated 1%Nd, 2.5-10%Th co-doped transparent ceramic and got laser output in 1974. The grain size could be kept around 120 μ m after sintering at 2170℃. Belyakov (13) et al introduced HfO$_2$ for the sintering of yttria, cation vacancies were formed in the inner parts of the crystal, which facilitates mass diffusion and perfects the structure inside the crystal. The segregation of hafnium oxide at the crystal boundaries leads to the formation of oxygen vacancies in the material, which impedes mass transfer between the crystals and facilitates their rapid growth and the capture of intercrystalline pores.

Rhodes (14) used oxalic acid co-precipitation method to produce lanthanum doped yttria powder and sintered by controlled transient solid second-phase sintering method. Materials with near theoretical total transmittances and specular transmittance within 6% of theoretical were obtained.

According to Y$_2$O$_3$-La$_2$O$_3$ binary phase diagram, limited solid solution can be produced when the mole ratio of La$_2$O$_3$ is less than 14%. As a trivalent dopant, La^{3+} could be dissolved into the yttria crystalline lattice uniformly and enhance the grain boundary mobility in yttria(15).. The appearance of La^{3+} could change the crystalline field of yttria and the relevant optical properties. Nd^{3+},La^{3+}:Y$_2$O$_3$ and Yb^{3+}, La^{3+}:Y$_2$O$_3$ transparent ceramics were fabricated in this paper.

2. EXPERIMENTAL

Highly purity Y$_2$O$_3$, La$_2$O$_3$, Yb$_2$O$_3$, and Nd$_2$O$_3$ powders were used as starting materials. We prepared highly sinterable Nd^{3+}, La^{3+}:Y$_2$O$_3$ and Yb^{3+}, La^{3+}:Y$_2$O$_3$ powders by ammonia precipitation method. The starting powders were dissolved in nitric acid to form a mixture of nitrate solutions. Then ammonia solution was added dropwise under stirring. After aging, the precursor gel was washed by deionized water for several times to remove the byproducts. After freeze-drying, the precursor was calcined at 1000° C. The powder was uniaxially compacted into pellets, followed by sintering in vacuum at 1750°C for 4h. The sintered samples were annealed in air at 1500 ° C for 10h. Then the sample was polished for further testing.

Mirror-polished samples were thermal-etched at 1600 for 20min. EPMA (Model JXA-8100, JEOL, Japan) was used to determine the grain size, at least 300 grains were measured to get the mean size. The grain sizes were measured by the linear-intercept method and calculated from G=1.5L, where G is the average grain size and L is the average intercept length. Mirror-polished samples on both surfaces were used to measure the optical transmittance (Model U-2800 Spectrophotometer, Hitachi, Japan). The fluorescence spectra and fluorescent lifetime of the ceramics were measured by a spectrofluorimeter (Fluorolog-3, Jobin Yvon, France).

3. RESULTS AND DISCUSSION

Fig.1(a) shows the microphotographs of 2at%Nd^{3+}, 7at%La^{3+}:Y$_2$O$_3$ sample, while Fig.1(b) shows the microphotographs of 3at% La, 2at%Nd co-doped yttria, both samples were sintered at 1750° C for 4h. Few pores could be found in the samples, and the mean grain size for 3at% La doping sample was 21±6μm, while for 7at% La doping sample was about 18±5μm, both value are smaller than pure yttria sintered at 1750° C in vacuum at same holding time shown in our experiments. It is concluded that the appearance of lanthanum ion may inhibit the grain growth through solute-drag mechanism during sintering. Solute-drag mechanism makes it possible to fully eliminate pores in the ceramics.

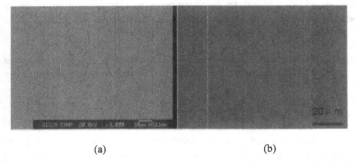

(a)   (b)

Fig.1 The microphotographs of 7at% La, 2at%Nd co-doped yttria and 3at% La, 2at%Nd co-doped yttria samples

The room temperature absorption band area of $^4I_{9/2} \rightarrow [^4F_{5/2}+^2H(2)^{9/2}]$ for 2at%Nd^{3+}, 7at%La^{3+}:Y₂O₃ is shown in Fig.2(a). The maximum absorption coefficient is at 821nm. And it is found that the absorption peak bands are broadened compared to Nd:Y₂O₃ samples. The full width at half maximum (FWHM) of the peak centered at 803nm is 10nm. The absorption cross section of the peak located 808nm is $2.75 \times 10^{-24}m^2$. The broadened absorption peak is benefit for lowering the request of pump devices. Fig.2(b) shows the room temperature emission spectra of the sample. The intermanifold branching ratio ($\beta_{JJ'}$) at 1078nm is 52.2%, which means most of the energy go down through the $^4F_{3/2}$ $\rightarrow^4I_{11/2}$ channel, which is similar to Nd:YAG. While in samples without La doping, $^4F_{3/2}\rightarrow^4I_{9/2}$ is the main channel. The emission cross section is about $2.10 \times 10^{-24}m^2$ at 1078nm. The fluorescence lifetime of the sample is 0.315ms.

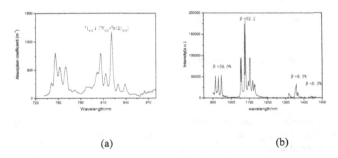

(a)   (b)

Fig.2 Room temperature absorption and emission spectra of Nd,La:Y₂O₃ ceramics.

Fig. 3 presents the absorption and emission spectra of 3% La, 2%Yb co-doped yttria transparent ceramic. There are three absorption peaks in Fig. 3(a), which are centered at 904, 948, and 974nm, all attributed to the $^2F_{7/2} \rightarrow ^2F_{5/2}$ transition of Yb^{3+} ion. The absorption spectra is similar to those of $Yb:Y_2O_3$. The broad absorption band is benefit for absorbing the pump energy, and it can lower the high request for temperature controlling. The highest absorption cross section is $4.13 \times 10^{-25} m^2$ at 978nm, and $3.62 \times 10^{-25} m^2$ at the LD pumped wavelength (940nm).

There are two main emission peaks between 980 and 1100nm wavelength in Fig. 3(b), centered at 1030nm and 1078nm, respectively. According to F-L equation, the emission cross sections are shown in Figure 3(b).

$$\sigma_{em}(\lambda) = \frac{\lambda^5}{8\pi n^2 c\tau} \times \frac{I(\lambda)}{\int \lambda I(\lambda) d\lambda}$$

Where $\sigma_{em}(\lambda)$ is the emission cross section at the wavelength, λ is the wavelength, n is the refractive index, c is velocity of light, τ is the $^2F_{5/2}$ fluorescent lifetime, $I(\lambda)$ is the intensity of fluorescent emission at the wavelength.

The highest emission cross section is $2.31 \times 10^{-24} m^2$ at 1030nm. This value is suitable for realizing laser oscillation. The full width at half maximum (FWHM) of the 1030nm emission peak is about 16nm, and 14nm at 1078nm emission peak, which is a potential candidate as femtosecond short pulse laser materials by the mode locking technologies. The fluorescent lifetime is 0.92ms, which is higher than that of yttria crystal. The long fluorescent lifetimes are attractive for energy storage and make it suitable for high power laser output.

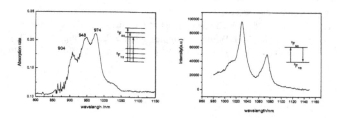

Fig.3 Room temperature absorption and emission spectra of Yb,La:Y₂O₃ ceramics.

3. CONCLUSIONS

1. It is easy to fabricate yttria transparent ceramics with La as the sintering additive. The grain growth is inhabited during the sintering process because of the solute drag mechanism;
2. The crystal structure is changed after La doping, the absorption band is broadened, and it is benefit to absorb more energy and reduce the request for pump.
3. Both Nd^{3+}, $La^{3+}:Y_2O_3$ and Yb^{3+}, $La^{3+}:Y_2O_3$ are promising candidate for laser application. After doping with La, Nd^{3+}, $La^{3+}:Y_2O_3$ samples have the similar emission channel to Nd:YAG..

ACKNOWLEDGEMENT
This work was supported by the Shanghai Science and Technology Committee (No. 07DJ14001) and the State Key Laboratory of High Performance Ceramics and Superfine Microstructures.

REFERENCES

1. Lu, J., Murai, T., Takaichi, K., Uematsu, T., and Ueda, K. (2001) Nd3+: Y 2 O 3 CERAMIC LASER, *Jpn. J. Appl. Phys., Part 2 40*.

2. Lu, J., Murai, T., Takaichi, K., Uematsu, T., Ueda, K., Yagi, H., Yanagitani, T., and Kaminskii, A. A. (2002) Highly efficient CWNd : Y2O3 ceramic laser, in *Topical Meeting on Advanced Solid-State Lasers* (Fermann, M. E. M. L. R., Ed.), pp 318-320, Quebec City, CANADA.

3. Brenier, A., and Boulon, G. (2001) Overview of the best Yb3+-doped laser crystals, *Journal of Alloys and Compounds 323*, 210-213.

4. Lupei, A., Lupei, V., Taira, T., Sato, Y., Ikesue, A., and Gheorghe, C. (2002) Energy transfer processes of Nd3+ in Y2O3 ceramic, in *International Conference on Luminescence and Optical Spectroscopy of Condensed Matter*, pp 72-76, Budapest, Hungary.

5. Ikesue, A., and Aung, Y. L. (2006) Synthesis and performance of advanced ceramic lasers, in *9th International Ceramic Processing Science Symposium*, pp 1936-1944, Coral Spring, FL.

6. Lupei, V., Lupei, A., and Ikesue, A. (2008) Transparent polycrystalline ceramic laser materials, *Optical Materials 30*, 1781-1786.

7. Lu, J., Takaichi, K., Uematsu, T., Shirakawa, A., Musha, M., Ueda, K., Yagi, H., Yanagitani, T., and Kaminskii, A. A. (2002) Promising ceramic laser material: Highly transparent Nd3+: Lu2O3 ceramic, *Applied Physics Letters 81*, 4324-4326.

8. Lefever, R. A., and Matsko, J. (1967) Transparent yttrium oxide ceramics, *Materials Research Bulletin 2*, 865-869.

9. Greskovich, C., and Oclair, C. R. (1986) Transparent, sintered Y2-xSrxO3-x/2 ceramics, pp 350-355, Advanced ceramic materials.

10. Ikegami, T., Li, J. G., Mori, T., and Moriyoshi, Y. (2002) Fabrication of transparent yttria ceramics by the low-temperature synthesis of yttrium hydroxide, *Journal of the American Ceramic Society 85*, 1725-1729.

11. Ikesue, A., Kamata, K., and Yoshida, K. (1996) Effects of neodymium concentration on optical characteristics of polycrystalline Nd:YAG laser materials, *Journal of the American Ceramic Society 79*, 1921-1926.

12. Greskovich, C., and Chernoch, J. P. (1974) Improved Polycrystalline Ceramic Lasers, *Journal of Applied Physics 45*, 4495-4502.

13. Belyakov, A. V., Lemeshev, D. O., Lukin, E. S., Val'nin, G. P., and Grinberg, E. E. (2006) Optically transparent ceramics based on yttrium oxide using carbonate and alkoxy precursors, *Glass and Ceramics 63*, 262-264.

14. Rhodes, W. H. (1981) Controlled Transient Solid Second-Phase Sintering of Yttria, *Journal of the American Ceramic Society 64*, 13-19.

15. Dou, C. G., Yang, Q. H., Hu, X. M., and Xu, J. (2008) Cooperative up-conversion luminescence of ytterbium doped yttrium lanthanum oxide transparent ceramic, *Optics Communications 281*, 692-695.

METAL OXIDE NANOELECTRODES FOR ENVIRONMENTAL SENSORS - ZNO RODS AND PARTICULATE FILMS

Yoshitake Masuda[1], Dewei Chu[1], Xiulan Hu[1], Tatsuki Ohji[1], Kazumi Kato[1], Masako Ajimi[2], Makoto Bekki[2] and Shuji Sonezaki[2]

1 National Institute of Advanced Industrial Science and Technology (AIST)

2266-98 Anagahora, Shimoshidami, Moriyama-ku, Nagoya 463-8560, Japan

*2 TOTO Ltd. Research Laboratory, 1-1, Nakashima 2-chome, Kokurakita-ku, Kitakyushu, Fukuoka 802-8601, Japan * Corresponding Author: Y. Masuda, masuda-y@aist.go.jp*

ABSTRACT

ZnO structures were fabricated on substrates for electrodes of dye-sensitized molecular sensors. They were crystallized in the aqueous solutions at 60 °C. Precise control of crystallization allowed us to form unique structures. ZnO rods were formed on fluorine doped tin oxide substrates without seed layers. They were 100-120 nm in diameter and 300-400 nm in length. They stood perpendicular to the substrates. Lengthwise direction of hexagonal prism was parallel to c-axis of ZnO crystal structure. They grew along long c-axis to form anisotropic shape. The ZnO rod arrays covered with dye using 1 μM and 100 nM solution and without dye showed photocurrent of 1.12×10^{-7} A, 5.75×10^{-9} A and 2.19×10^{-9} A, respectively, under irradiation using 0.7 mW lasers. A high signal-to-noise ratio (S/N) of about 50 (=1.12×10^{-7} A / 2.19×10^{-9} A) was obtained under the irradiation. Moreover, ZnO particulate films were prepared on glass substrates in the solutions. The films consisted of multi-needle ZnO particles and thin sheets. The particles were firstly prepared at 60 °C. The thin sheets were then formed in the solutions at 25 °C. Thin sheets had a thickness of 10 − 50 nm and width of 1 − 10 μm and were connected to particles closely with no clearance. The particulate films had continuous open pores ranging from several nm to 10 μm in diameter.

INTRODUCTION

Unique structures of ZnO have been synthesized in various morphologies like nanorods, nanowires, nanobelts, nanorings, nanohelices and so on. They had high surface area and high crystallinity. The wurtize structure of typical ZnO favors the formation of anisotropic structures. ZnO is a candidate material for various sensors[1-4] and dye-sensitized solar cells[5-9]. High surface area, high conductivity and high crystallinity are strongly required for these devices. ZnO structures have been applied for electrodes of sensor and solar cells. Aqueous solution systems were developed to synthesize ZnO structures. Hexagonal symmetry radial whiskers[10], cylindrical hexagonal rods[11], ellipses[11], or multineedle shapes[11] were, for instance, prepared in the solutions. Patterning of them was realized on the substrates using surface interaction between ZnO and self-assembled monolayers[11]. Aqueous solution systems have important advantages in ordinary temperature process, low cost, environmentally friendly process, simple process etc. The system can be applied to metal oxide device fabrication on polymer films. They can be combined with biotechnology and organic chemistry.

In this study, ZnO structures were fabricated on substrates for electrodes of dye-sensitized molecular sensors. They were crystallized in the aqueous solutions at 60 °C. Precise control of crystallization allowed us to form unique structures. Morphology and crystal structure of them were investigated in details. Their deposition mechanism was discussed.

EXPERIMENT

Synthesis of ZnO

Zinc nitrate hexahydrate ($Zn(NO_3)_2 \cdot 6H_2O$, > 99.0%, MW 297.49, Kanto Chemical Co., Inc.) (15 mM) was dissolved in distilled water at 60°C. Ethylenediamine ($H_2NCH_2CH_2NH_2$, > 99.0%, MW 60.10, Kanto Chemical Co., Inc.) (15 mM) was added to the solutions to induce the formation of ZnO.

Glass substrates coated with FTO transparent conductive films (FTO, SnO_2: F, Asahi glass Co., Ltd., 9.3-9.7 $\Omega/\square$, 26 × 50 × 1.1 mm) were exposed to ultraviolet light (low-pressure mercury lamp PL16-110) for 10 min. The initial FTO substrate showed a water contact angle of 96°. The UV-irradiated surfaces were, however, wetted completely (contact angle 0–1°). The FTO substrates were modified to super-hydrophilic surfaces. The FTO substrates were immersed in the middle of the solutions with the bottom up, tilted at 15 degrees to the upright. The FTO substrates were kept for 6 h with no stirring.

ZnO particulate films constructed from ZnO particles and thin sheets were fabricated by immersion for 48 h. Glass substrates were immersed in the middle of the solutions at an angle and the solutions were kept at 60 °C using a water bath for 6 h with no stirring. The solutions were then left to cool for 42 h in the bath. The solutions became clouded shortly after the addition of ethylenediamine and clear after 6 h. The bottom of the solutions were covered with white precipitate after 6 h.

Characterizations

Morphologies of ZnO particles were observed with a field emission scanning electron microscope (FE-SEM; JSM-6335FM, JEOL Ltd.). The crystal phases were evaluated with an X-ray diffractometer (XRD; RINT-2100V, Rigaku) with CuKα radiation (40 kV, 40 mA).

The photocurrent action spectra were investigated after dye adsorption described as follows. The 1 µM or 100 nM of ssDNA-Cy5 (cy5-DP53-t: Cy5-GCGGCATGAACCTGAGGCCCATCCT) solution was dropped with pipettes onto the substrates[3]. For comparison, ssDNA (lambda-gt10: TTGAGCAAGTTCAGCCTGGTTAAG) solution was dropped onto the substrates. The ssDNA solution contained no dye. The substrates were dried at 95°C for 10 min in air. They were rinsed 3 times in sodium dodecyl sulfate (SDS, $NaC_{12}H_{25}SO_4$) for 15 min each time and rinsed 3 times in ultrapure water. They were then boiled in water for 2 min, immersed in dehydrated ethanol at 4°C for 1 min and dried by strong air flow. The photoelectrochemical experiments were performed in sandwich-type cells, which were constructed using DNA labeled with dye-modified ZnO electrodes (working electrodes), Pt counterelectrodes, and electrolyte solutions containing the redox pair (I^-/I_3^-) to fill the cells. Two types of red lasers (low intensity 0.7mW with wavelength of 632 nm and high intensity 97 mW with wavelength of 650 nm) served as light sources. The photocurrent actions were determined by measuring the short-circuit photocurrents at excitation wavelengths of red lasers. All the measurements were conducted at room temperature.

RESULTS AND DISCUSSION

(1) ZnO rods

ZnO rods were formed on FTO substrates in the solutions (Fig. 1a). It is notable that the rod arrays were prepared on the substrates without seed layers and high temperature annealing. Additionally, the arrays can be prepared on low heat resistant substrates such as conductive polymer films for light weight flexible devices which can be bended. About 160 rods were observed per 10 μm^2 on FTO substrates. They have hexagonal prism shapes and were about 100 nm in diameter. The distance between the rods ranges between 100 and 1000 nm. Cross-sectional images showed uniform formation of the rods on whole areas of FTO substrates (Fig. 1b). They were 100-120nm in diameter and 300-400 nm in length. Aspect ratio was estimated to 3-4. The rods stood perpendicular to the substrates with variation of about 20 °. The variation was mainly caused by uneven surfaces of FTO layers. Magnified images of the rods showed V-shaped tips and flat side crystal faces of hexagonal prisms.

The substrates showed X-ray diffraction peaks at 2θ = 26.5, 33.8, 37.9, 42.6, 51.6, 54.6, 61.7, 65.8, 71.0, 78.6, 80.9, 89.5, 90.6 and 93.0 ° (Fig. 2). They were assigned to crystalline SnO_2, which stemmed from the FTO surface coatings on glass substrates. The substrates also showed an X-day diffraction peak at 2θ = 34.4 °. It was assigned to 0002 X-ray diffraction peak of hexagonal wurtzite ZnO (JCPDS No. 36-1451). The arrays showed only 0002 diffraction peak though randomly deposited ZnO usually had high diffraction intensities from (10-10) and (10-11) planes compared to that from (0002) planes. It indicated a high c-axis orientation of the arrays. ZnO rods had many stacks of (0002) planes which were parallel to the substrates and small amount of stacks of (10-10) and (10-11) planes due to the anisotropic crystal growth along c-axis.

(2) Photo-electric effect of dye-sensitized ZnO rods

Photocurrents from ZnO rod arrays covered with dye-labeled DNA were evaluated under irradiation of red lasers. Electrons excited by the lasers were injected into ZnO rods. They transferred through trap sites of ZnO by the electron hopping process. These electrons diffused to counter electrodes, and reacted with trapped holes in the oxidized dyes via the redox pair in the electrolyte solutions without decomposing the dye itself.

ZnO rod arrays covered with dye using 1 μM and 100 nM solution had photocurrent of 7.37 × 10^{-6} A and 5.50 × 10^{-7} A, respectively, under irradiation using 97 mW lasers. The ZnO rod arrays without dye, on the other hand, showed a rather low photocurrent of 2.80 × 10^{-7} A. ZnO rods showed high S/N ratio of about 30 (= 7.37 × 10^{-6} A / 2.80 × 10^{-7} A) under the irradiation. The ZnO rod arrays covered with dye using 1 μM and 100 nM solution and without dye showed photocurrent of 1.12×10^{-7} A, 5.75×10^{-9} A and 2.19×10^{-9} A, respectively, under irradiation using 0.7 mW lasers. A high signal-to-noise ratio (S/N) of about 50 (=1.12×10^{-7} A / 2.19×10^{-9} A) was obtained under the irradiation. The noise current was usually detected from ZnO without dye under irradiation by excitation lasers. It decreases sensitivity of the sensors. Therefore, low noise current and high S/N ratios are strongly required for high sensitive sensors. High S/N ratios caused from low noise currents are advantages in this system to apply them for high sensitive sensors.

Grain boundaries in oxide electrodes are known to decrease electrical conductivity in sensors and solar cells. In contrast, the ZnO rod arrays developed in this study consisted of single crystals which have no grain boundaries. Additionally, ZnO rods grew on FTO substrates directly, have no intermediate layers as shown in TEM observations. These were effective to transport photo-generated electrons to the FTO electrodes. Low noise current and high S/N ratios were thus successfully obtained in this system.

(3) ZnO particulate films

ZnO particulate films consisted of multi-needle particles and thin sheets (Fig. 3). The particles had many needles that grew from the center of the particles. Needles were constructed from an assembly of narrow acicular crystals. The side surfaces of needles were covered with arrays of pleats. The tips of the needles were rounded V-shape with many asperities. Edged hexagonal shapes were observed at the tips of needles. The c-axis would be the lengthwise direction of multi-needles and narrow acicular crystals. Thin sheets had a thickness of 10 – 50 nm and width of 1 – 10 μm and were connected to particles closely with no clearance. The particulate films had continuous open pores ranging from several nm to 10 μm in diameter. The particulate films showed X-ray diffraction patterns of ZnO crystal with no additional phase (Fig. 4). Diffraction peaks were very sharp, showing high crystallinity of the particulate films. The high intensity of 0002 would be caused by elongation of multi-needle particles in the c-axis direction which increases the stacks of (0002) crystal planes.

ZnO multi-needle particles having an ultrafine surface relief structure were prepared at 60 °C in the white solution during the initial 80 min. The supersaturation degree was high at the initial stage of the reaction due to the high concentration of ions. ZnO particles were then precipitated, making the bottom of the solution white and the solution itself clear. Ions were consumed to form ZnO particles and thus the ion concentration of the solution decreased rapidly. Thin sheets were formed at 25 °C in the clear solution after the formation of multi-needle particles. Solution temperature and supersaturation degree would influence on the morphology of precipitates. Consequently, the particulate films constructed from multi-needle particles and thin sheets were successfully fabricated by the two-step growth.

Thin sheets were transformed to particles and porous particulate films by annealing at 500 °C for 1 h in air. The sheets did not maintain their thin sheet shape due to high slimness and/or phase transformation. Thin sheets would be inorganic films containing Zn ions such as crystalline ZnO, amorphous ZnO or zinc hydroxide and were transformed to porous ZnO particulate films by the annealing. Further investigation of the thin sheets would contribute to more precise morphology control of ZnO structure and further improvement of specific surface area. Additionally, precise evaluation of mechanical strength and electrical properties should be performed to clarify the potential of ZnO particulate films for sensors or solar cells and to produce guidelines for improving their properties.

CONCLUSION

Morphology control of ZnO was realized in aqueous solutions. ZnO rods were fabricated on FTO substrates for electrodes of dye-sensitized molecular sensors. They were 100-120 nm in diameter and 300-400 nm in length. They grew along long c-axis to form anisotropic shape. ZnO rods covered with dye using 1 μM and 100 nM solution had photocurrent of 7.37×10^{-6} A and 5.50×10^{-7} A, respectively, under irradiation using 97 mW lasers. The ZnO rod arrays without dye, on the other hand, showed a rather low photocurrent of 2.80×10^{-7} A. ZnO rods showed high S/N ratio of about 30 (= 7.37×10^{-6} A / 2.80×10^{-7} A) under the irradiation. The ZnO rod arrays covered with dye using 1 μM and 100 nM solution and without dye showed photocurrent of 1.12×10^{-7} A, 5.75×10^{-9} A and 2.19×10^{-9} A, respectively, under 0.7 mW lasers. A high signal-to-noise ratio (S/N) of about 50 (=1.12×10^{-7} A / 2.19×10^{-9} A) was obtained under the lower power irradiation. Moreover, ZnO particulate films were prepared on glass substrates in the solutions. The films consisted of multi-needle ZnO particles and thin sheets. The particles were firstly prepared at 60 °C. The thin sheets were then formed in the solutions at 25 °C. Thin sheets had a thickness of 10 – 50 nm and width of 1 – 10 μm and were connected to particles closely with no clearance. The particulate films had continuous open pores ranging from several nm to 10 μm in diameter.

ACKNOWLEDGEMENT
This work was supported by METI, Japan, as part of R&D for High Sensitivity Environment Sensor Components.

REFERENCES

[1] N. Kumar, A. Dorfman, and J. Hahm, "Ultrasensitive DNA sequence detection using nanoscale ZnO sensor arrays," Nanotechnology, 17(12), 2875-81 (2006).

[2] G. Ramsay, "DNA chips: State-of-the-art," Nature Biotechnology, 16(1), 40-44 (1998).

[3] H. Tokudome, Y. Yamada, S. Sonezaki, H. Ishikawa, M. Bekki, K. Kanehira, and M. Miyauchi, "Photoelectrochemical deoxyribonucleic acid sensing on a nanostructured TiO2 electrode," Appl. Phys. Lett., 87(21), 213901-03 (2005).

[4] Y. Masuda and K. Kato, "Rapid growth of thick particulate film of crystalline ZnO in an aqueous solution," Thin Solid Films, 516, 2474-77 (2008).

[5] M. Law, L. E. Greene, J. C. Johnson, R. Saykally, and P. D. Yang, "Nanowire dye-sensitized solar cells," Nature Mater., 4(6), 455-59 (2005).

[6] J. B. Baxter and E. S. Aydil, "Nanowire-based dye-sensitized solar cells," Appl. Phys. Lett., 86(5), 53114 (2005).

[7] R. Katoh, A. Furube, K. Hara, S. Murata, H. Sugihara, H. Arakawa, and M. Tachiya, "Efficiencies of electron injection from excited sensitizer dyes to nanocrystalline ZnO films as studied by near-IR optical absorption of injected electrons," J. Phys. Chem. B, 106(50), 12957-64 (2002).

[8] S. Karuppuchamy, K. Nonomura, T. Yoshida, T. Sugiura, and H. Minoura, "Cathodic electrodeposition of oxide semiconductor thin films and their application to dye-sensitized solar cells," Solid State Ionics, 151(1-4), 19-27 (2002).

[9] K. Keis, C. Bauer, G. Boschloo, A. Hagfeldt, K. Westermark, H. Rensmo, and H. Siegbahn, "Nanostructured ZnO electrodes for dye-sensitized solar cell applications," J. Photochem. Photobiol. A, 148(1-3), 57-64 (2002).

[10] Y. Masuda, N. Kinoshita, and K. Koumoto, "Hexagonal Symmetry Radial Whiskers of ZnO Crystallized in Aqueous Solution," J. Nanosci. Nanotechnol., 9, 522-26 (2009).
[11] Y. Masuda, N. Kinoshita, F. Sato, and K. Koumoto, "Site-selective deposition and morphology control of UV- and visible-light-emitting ZnO crystals," Cryst. Growth Des., 6(1), 75-78 (2006).

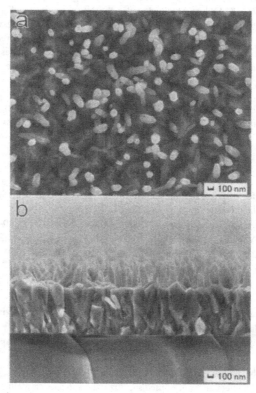

Figure 1. SEM micrographs of ZnO rod array on a FTO substrate. (a) cross-section (b) top view.

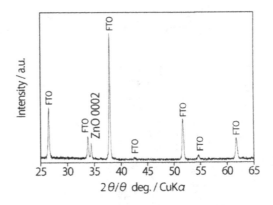

Figure 2. XRD pattern of ZnO rod array on a FTO substrate.

Figure 3. SEM micrographs of particulate film constructed of ZnO multi-needle particles and thin sheets.

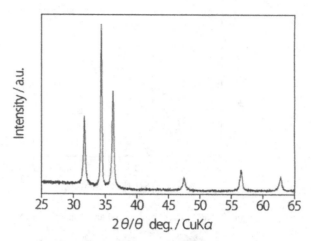

Figure 4. XRD diffraction pattern of particulate film constructed of ZnO multi-needle particles and thin sheets.

Author Index

Printed in the United States
By Bookmasters